Applied EPR Spectroscopy

Applied EPR Spectroscopy

Editor

Yordanka Karakirova

Basel • Beijing • Wuhan • Barcelona • Belgrade • Novi Sad • Cluj • Manchester

Editor
Yordanka Karakirova
Institute of Catalysis
Bulgarian Academy of Sciences
Sofia
Bulgaria

Editorial Office
MDPI
St. Alban-Anlage 66
4052 Basel, Switzerland

This is a reprint of articles from the Special Issue published online in the open access journal *Molecules* (ISSN 1420-3049) (available at: www.mdpi.com/journal/molecules/special_issues/applied_EPR_spectroscopy).

For citation purposes, cite each article independently as indicated on the article page online and as indicated below:

Lastname, A.A.; Lastname, B.B. Article Title. *Journal Name* **Year**, *Volume Number*, Page Range.

ISBN 978-3-7258-0350-7 (Hbk)
ISBN 978-3-7258-0349-1 (PDF)
doi.org/10.3390/books978-3-7258-0349-1

© 2024 by the authors. Articles in this book are Open Access and distributed under the Creative Commons Attribution (CC BY) license. The book as a whole is distributed by MDPI under the terms and conditions of the Creative Commons Attribution-NonCommercial-NoDerivs (CC BY-NC-ND) license.

Contents

About the Editor vii

Preface ix

Elizaveta Kobeleva, Ekaterina Shabratova, Adi Azoulay, Rowan W. MacQueen, Neeta Karjule and Menny Shalom et al.
Long-Term Characterization of Oxidation Processes in Graphitic Carbon Nitride Photocatalyst Materials via Electron Paramagnetic Resonance Spectroscopy
Reprinted from: *Molecules* **2023**, *28*, 6475, doi:10.3390/molecules28186475 1

Bogumił Cieniek, Ireneusz Stefaniuk, Ihor Virt, Roman V. Gamernyk and Iwona Rogalska
Zinc–Cobalt Oxide Thin Films: High Curie Temperature Studied by Electron Magnetic Resonance
Reprinted from: *Molecules* **2022**, *27*, 8500, doi:10.3390/molecules27238500 16

Aleksandra Wedrychowicz, Bogumił Cieniek, Ireneusz Stefaniuk, Ihor Virt and Romana Śliwa
Electron Paramagnetic Resonance Study of PbSe, PbTe, and PbTe:In Semiconductors Obtained by the Pulsed Laser Deposition Method
Reprinted from: *Molecules* **2022**, *27*, 4381, doi:10.3390/molecules27144381 27

Krisztina Sebők-Nagy, Zoltán Kóta, András Kincses, Ákos Ferenc Fazekas, András Dér and Zsuzsanna László et al.
Spin-Label Electron Paramagnetic Resonance Spectroscopy Reveals Effects of Wastewater Filter Membrane Coated with Titanium Dioxide Nanoparticles on Bovine Serum Albumin
Reprinted from: *Molecules* **2023**, *28*, 6750, doi:10.3390/molecules28196750 39

Aygun Nasibova, Rovshan Khalilov, Mahammad Bayramov, İslam Mustafayev, Aziz Eftekhari and Mirheydar Abbasov et al.
Electron Paramagnetic Resonance Studies of Irradiated Grape Snails (*Helix pomatia*) and Investigation of Biophysical Parameters
Reprinted from: *Molecules* **2023**, *28*, 1872, doi:10.3390/molecules28041872 56

Antonino Famulari, Danilo Correddu, Giovanna Di Nardo, Gianfranco Gilardi, George Mitrikas and Mario Chiesa et al.
Heme Spin Distribution in the Substrate-Free and Inhibited Novel CYP116B5hd: A Multifrequency Hyperfine Sublevel Correlation (HYSCORE) Study
Reprinted from: *Molecules* **2024**, *29*, 518, doi:10.3390/molecules29020518 69

Zuzanna Kabacińska and Alida Timar-Gabor
Dating Sediments by EPR Using Al-h Centre: A Comparison between the Properties of Fine (4–11 μm) and Coarse (>63 μm) Quartz Grains
Reprinted from: *Molecules* **2022**, *27*, 2683, doi:10.3390/molecules27092683 91

Amanda Burg Rech, Angela Kinoshita, Paulo Marcos Donate, Otaciro Rangel Nascimento and Oswaldo Baffa
Electron Spin Resonance Dosimetry Studies of Irradiated Sulfite Salts
Reprinted from: *Molecules* **2022**, *27*, 7047, doi:10.3390/molecules27207047 109

Yordanka Karakirova
Application of Amino Acids for High-Dosage Measurements with Electron Paramagnetic Resonance Spectroscopy
Reprinted from: *Molecules* **2023**, *28*, 1745, doi:10.3390/molecules28041745 118

Wilfred R. Hagen
Conversion of a Single-Frequency X-Band EPR Spectrometer into a Broadband Multi-Frequency 0.1–18 GHz Instrument for Analysis of Complex Molecular Spin Hamiltonians
Reprinted from: *Molecules* **2023**, *28*, 5281, doi:10.3390/molecules28135281 . . . . **131**

Vasily V. Ptushenko and Vladimir N. Linev
A Review of the Dawn of Benchtop EPR Spectrometers—Innovation That Shaped the Future of This Technology
Reprinted from: *Molecules* **2022**, *27*, 5996, doi:10.3390/molecules27185996 . . . . **147**

About the Editor

Yordanka Karakirova

Dr. Yordanka Karakirova is an Associate Professor at the Institute of Catalysis, BAS. She completed her higher education with the Faculty of Chemistry and Pharmacy at the Sofia University in 2005. Her PhD thesis was received by the Institute of Catalysis, BAS, in 2008. Her main activities and scientific interests are related to electron paramagnetic resonance (EPR) spectroscopy, UV spectrophotometry, quantitative EPR spectroscopy, EPR dosimetry, the EPR investigation of irradiated foods, free radicals, saccharides, and solid state chemistry. She is the Head of the Center of EPR spectroscopy at the Institute of Catalysis, the Chairman of the Bulgarian EPR Society, and members of the European Federation of EPR Societies and the Bulgarian Catalytic Society. In 2016, she was awarded the National Scholarship of Loreal Bulgaria by the program "For the Women in Science" for her project entitled "Development of new type solid state/EPR dosimeters for high energy radiation".

Preface

Electron paramagnetic resonance (EPR) spectroscopy has been widely used for more than 60 years, mainly due to its advantages, such as its high sensitivity (to 10^{12}–10^{13} molecules, which translates to 10^{-11} to 10^{-12} M) and selectivity (only for paramagnetic species). Moreover, it is a quick, non-destructive method, and in some cases, samples may be kept as documents for future inspection. These basic qualities of X-band EPR spectroscopy have been enhanced with new generations of spectroscopies, which have introduced advantages such as the ability to use multifrequency approaches, improved time resolutions, better pulsed techniques, and the incorporation of multiple irradiations. These developments have significantly extended the range of applications of EPR spectroscopy to different fields, including EPR research in biology and medicine, chemistry (metal complexes, polymers, and catalysis), dosimetry, geology, mineralogy, archaeology, environmental control, food technology, quantitative EPR, etc. The design of portable EPR spectrometers with small dimensions greatly facilitates the use of this method in practice. The aim of the Special Issue "Applied EPR spectroscopy" is to highlight the current status as well as trends in the development and application of EPR spectroscopy.

Additionally, I want to thank all the authors and co-authors of this book who contributed to this Special issue.

Yordanka Karakirova
Editor

Article

Long-Term Characterization of Oxidation Processes in Graphitic Carbon Nitride Photocatalyst Materials via Electron Paramagnetic Resonance Spectroscopy

Elizaveta Kobeleva [1], Ekaterina Shabratova [1], Adi Azoulay [2], Rowan W. MacQueen [1], Neeta Karjule [2], Menny Shalom [2], Klaus Lips [1,3,*] and Joseph E. McPeak [1,*]

[1] Berlin Joint EPR Laboratory and EPR4Energy, Department Spins in Energy Conversion and Quantum Information Science (ASPIN), Helmholtz-Zentrum Berlin für Materialien und Energie GmbH, Hahn-Meitner-Platz 1, 14109 Berlin, Germany
[2] Department of Chemistry, Ilse Katz Institute for Nanoscale Science and Technology, Ben-Gurion University of the Negev, Beer-Sheva 8410501, Israel
[3] Berlin Joint EPR Laboratory, Fachbereich Physik, Freie Universität Berlin, 14195 Berlin, Germany
* Correspondence: lips@helmholtz-berlin.de (K.L.); joseph.mcpeak@helmholtz-berlin.de (J.E.M.)

Citation: Kobeleva, E.; Shabratova, E.; Azoulay, A.; MacQueen, R.W.; Karjule, N.; Shalom, M.; Lips, K.; McPeak, J.E. Long-Term Characterization of Oxidation Processes in Graphitic Carbon Nitride Photocatalyst Materials via Electron Paramagnetic Resonance Spectroscopy. *Molecules* 2023, *28*, 6475. https://doi.org/10.3390/molecules28186475

Academic Editor: Yordanka Karakirova

Received: 29 July 2023
Revised: 31 August 2023
Accepted: 1 September 2023
Published: 6 September 2023

Copyright: © 2023 by the authors. Licensee MDPI, Basel, Switzerland. This article is an open access article distributed under the terms and conditions of the Creative Commons Attribution (CC BY) license (https://creativecommons.org/licenses/by/4.0/).

Abstract: Graphitic carbon nitride (gCN) materials have been shown to efficiently perform light-induced water splitting, carbon dioxide reduction, and environmental remediation in a cost-effective way. However, gCN suffers from rapid charge-carrier recombination, inefficient light absorption, and poor long-term stability which greatly hinders photocatalytic performance. To determine the underlying catalytic mechanisms and overall contributions that will improve performance, the electronic structure of gCN materials has been investigated using electron paramagnetic resonance (EPR) spectroscopy. Through lineshape analysis and relaxation behavior, evidence of two independent spin species were determined to be present in catalytically active gCN materials. These two contributions to the total lineshape respond independently to light exposure such that the previously established catalytically active spin system remains responsive while the newly observed, superimposed EPR signal is not increased during exposure to light. The time dependence of these two peaks present in gCN EPR spectra recorded sequentially in air over several months demonstrates a steady change in the electronic structure of the gCN framework over time. This light-independent, slowly evolving additional spin center is demonstrated to be the result of oxidative processes occurring as a result of exposure to the environment and is confirmed by forced oxidation experiments. This oxidized gCN exhibits lower H_2 production rates and indicates quenching of the overall gCN catalytic activity over longer reaction times. A general model for the newly generated spin centers is given and strategies for the alleviation of oxidative products within the gCN framework are discussed in the context of improving photocatalytic activity over extended durations as required for future functional photocatalytic device development.

Keywords: EPR spectroscopy; photocatalysis; carbon nitride; electronic relaxation; semiconductor; oxidation processes; power saturation analysis; light-dependent EPR; electronic structure characterization; graphitic materials

1. Introduction

In the context of the global search for new sustainable energy technologies, graphitic carbon nitride (gCN) has gained a lot of attention in the research community as a potentially stable, readily available, and non-toxic photocatalyst for H_2 production via water splitting and other energy dense molecular syntheses via CO_2 reduction reactions as well as for environmental remediation via pollutant destruction [1,2]. The success of these proof-of-concept reactions using gCN photocatalysts has prompted research towards the highest possible catalytic output through a variety of surface area enhancements, doping schemes, and other unique applications of graphene-like structural modifications [2] to target the shortcomings of gCN materials. However, the lack of conclusive theoretical explanations or experimental investigations into the understanding of the underlying physical processes restricts gCN photocatalytic efficiency to below industrially applicable values [1]. To date, numerous modifications of gCN structure have been reported to increase catalytic activity without a clear mechanistic reasoning as to how or what process contributes to these increased rates of production. Additionally, gCN continues to suffer from fast quenching of photocatalytic activity and continues to require platinum cocatalysts to improve the stability and reproducibility of catalytic output, preventing gCN from being employed on any large scale as a truly metal-free photocatalyst [3].

Electron paramagnetic resonance (EPR) spectroscopy is particularly well suited to study the electronic structure of gCN materials owing to its non-destructive methods of interrogation and high specificity to paramagnetic electrons. Together with other methods, EPR has been employed to directly monitor spin transitions of photocatalytic species during light-irradiation [2,4]. Figure 1a–c shows an overview of the structural and electronic properties of gCN. The tri-s-triazine units are comprised of alternating carbon and nitrogen atoms, where the electron density of gCN materials allows for electrons in the sp^2 orbitals of the carbon and nitrogen atoms to form a π-conjugated framework that extends over the entire triazine unit (Figure 1b, *green*). Unpaired electrons distributed over the π-conjugated structure are responsible for photocatalytic activity via transfer from the valence band (σ-type bonds, Figure 1b,c, *grey*) to the conduction band (π-conjugated structure, Figure 1b,c, *green*) under light irradiation [4]. Electrons in the π-conjugated structure are paramagnetic and are therefore observable by EPR spectroscopy. It has been shown that the EPR signal is observable in the dark state, presumably because electrons are already present in the π-conjugated structure; however, under light illumination the EPR double integral intensity increases indicating an increase in the number of spins present [4]. Each triazine unit acts as a contributor to photocatalytic activity and as such, it has been shown in numerous reports that catalytic activity increases with increasing surface area, which could be modified through various synthesis routes [5]. The stability of these materials has previously only been investigated on the scale of hours and a decrease in the efficiency of hydrogen evolution over multiple catalytic cycles has been observed [6]. This timescale is insufficient for commercial applications and therefore the stability should be evaluated over extended durations.

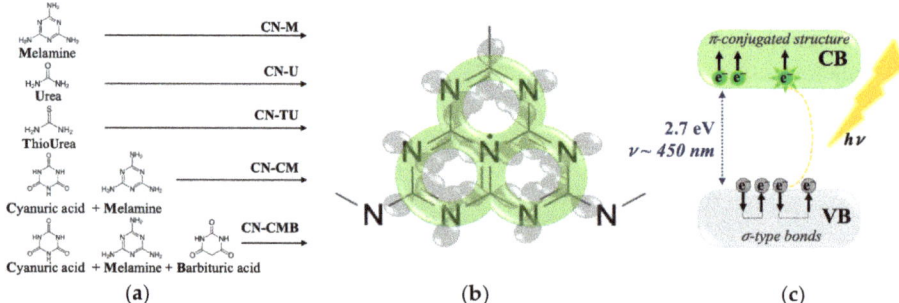

Figure 1. An overview of the structural and electronic properties of gCN. (**a**) Overview of precursors used to synthesize gCN materials; see methods for synthesis procedures. (**b**) The tri-s-triazine unit consists of alternating carbon and nitrogen atoms with bridging nitrogens connecting multiple units into a polymeric structure where single-unit π-conjugation (*green*) and σ-type bonds (*grey*) are shown. (**c**) Schematic illustration of the photoexcitation mechanism of gCN where electrons in the σ-type bonds, which are paired and therefore diamagnetic due to counteracting spin magnetic moments, are excited to the π-conjugated structure where they are then paramagnetic and observable by EPR [4].

Within this work the gCN electronic structure is further investigated by EPR spectroscopy to describe the current photocatalytic performance of gCN materials and to provide guidelines for further improvement of its photocatalytic efficiency. By providing direct and non-destructive access to the behavior of electrons in materials, examination of the dynamic properties of gCN photocatalysts is made possible in a way that allows increased understanding of the slow alterations to the gCN framework that affect long term stability. This investigation of gCN samples synthesized from various precursors additionally allows for the understanding of correlations between electronic structure and photocatalytic performance with respect to minute changes in synthesis procedures. In this study, we investigate graphitic carbon nitride (gCN) derived from various precursors, namely melamine (CN-M), urea (CN-U), thiourea (CN-TU), as well as supramolecular assemblies of cyanuric acid-melamine (CN-CM) and supramolecular assemblies of cyanuric acid-melamine-barbituric acid (CN-CMB) (Figure 1a). We herein present EPR investigations of gCN materials as a function of time, with saturation and relaxation behavior determined using a combination of both continuous wave and pulse EPR methods to define and probe the environment of the spins and underlying dynamic properties resulting in catalytic activity in gCN materials. Incorporation of laser excitation to classical CW-EPR experiments was similarly employed to study any effects on light response of the samples and to confirm the correlation between light-irradiation and EPR signal as previously reported [4,7]. Interactions with the latent environment were taken into consideration and were further evaluated using forced-oxidation schemes. To evaluate the impact of the herein reported changes in the EPR spectra with respect to time, the catalytic activity via hydrogen production rate measurements were performed using freshly synthesized gCN materials and subsequently reevaluated after twenty months of continuous exposure to normal atmosphere.

2. Results

2.1. gCN Electronic Stability over Time

The gCN EPR response has been mainly reported to be a single Lorentzian line and associated with unpaired electrons from π-conjugated structures responsible for photocatalytic activity [2,4,5,8]. More recent studies have shown evidence that multiple spins may contribute to the overall CW-EPR spectrum; however, variety in synthesis procedures leading to differences in morphology [5] does not allow for direct comparison of these results [7,9]. In this study EPR spectra were recorded initially and sequentially for 18 months following the initial synthesis of several gCN samples. Significant changes in the EPR spectrum were observed with respect to time and a deviation from the previously reported Lorentzian lineshape was observed in all gCN samples investigated.

Spectra recorded for all gCN samples investigated exhibited initially a more symmetric lineshape in agreement with previously reported observations [2,4]. As early as six months following the successful synthesis, the overall EPR lineshape was observed to deviate from a single Lorentzian for all five samples in the recorded spectra (Figure 2, *black*). The overall lineshape continued to change over time, providing initial evidence that a change in the electronic structure of the gCN framework has occurred. After 16 months following the successful gCN synthesis, an additional peak may be clearly distinguished (Figure 2, *red*). This observation suggests that an additional paramagnetic species arises in gCN over time and likely alters both the electronic structure of the catalytic material and the physical structure of the bonding character within the triazine units.

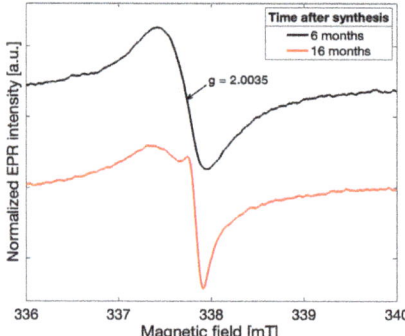

Figure 2. The CW-EPR spectrum recorded after 6 months (*black*) and 16 months (*red*) following successful synthesis of gCN (CN-M) under the same operating conditions. A deviation from a purely Lorentzian lineshape is observed near g = 2.0035 ± 0.00003, as shown by the black arrow in the spectra.

A fitting routine which considers two independent spin contributions to the overall spectrum was introduced using a linear combination of two derivative Lorentzian lineshapes according to the following formula,

$$I = m_1 \left(\frac{\Delta B_1^2}{(B-B_{o_1})^2 + \Delta B_1^2} \right)' + m_2 \left(\frac{\Delta B_2^2}{(B-B_{o_2})^2 + \Delta B_2^2} \right)'$$
$$= m_1 \frac{1}{\pi} \frac{16(B-B_{o_1})\Delta B_1}{\left(4(B-B_{o_1})^2 + \Delta B_1^2\right)^2} + m_2 \frac{1}{\pi} \frac{16(B-B_{o_2})\Delta B_2}{\left(4(B-B_{o_2})^2 + \Delta B_2^2\right)^2} \qquad (1)$$

where weighting coefficients m_1, m_2, central positions B_{o_1}, B_{o_2} and FWHM (full width at half maximum) ΔB_1, ΔB_2 values were allowed to vary within a least squares regression (performed in Matlab, *Mathworks*) [10]. In Figure 3a, the gCN spectrum shown previously in Figure 2 (*red*) is modelled using the above relationship. The overall fit is the sum of two Lorentzian lineshapes such that each independent Lorentzian lineshape can be built using one set of coefficients [m_1, B_{o_1}, ΔB_1] or [m_2, B_{o_2}, ΔB_2] obtained from the fit. The same fitting routine and initial parameters were utilized for all five investigated samples. Spectra recorded at different observation times and different microwave powers were found to be in good agreement throughout all samples. While the observed g-values of the two peaks were constant, the intensities and linewidths varied with microwave power, time, and temperature.

Here, the wider lineshape (Figure 3a, *green*, ΔB_{pp} = 0.6 mT) labeled as spin species 1, corresponds to unpaired electrons distributed over the π-conjugated structure of the triazine unit of gCN, also known to be responsible for photocatalytic activity, as both linewidth and g-value agree well with previous reports [2,4]. The origin of the narrow lineshape (Figure 3a, *blue*, ΔB_{pp} = 0.1 mT), labeled spin species 2, has not been investigated before.

By separating the two Lorentzian contributions to the overall EPR spectrum via the fitting routine described above, the peak-to-peak intensities of each species were determined for spectra recorded under the same conditions at different time points between 2 and 18 months after synthesis (Figure 3b). The intensity of the line associated with spin species 2 (Figure 3b, *blue*) increases drastically over time, while the intensity of the line associated with spin species 1, typically attributed to catalytically active spins, (Figure 3b, *green*) decreases slowly over the same time duration. The exponential increase in the intensity at g = 2.003 ± 0.00003 and slow decrease in the intensity at g = 2.004 ± 0.00003 provides insight into the dynamic nature of gCN. It is likely that the second spin species (g = 2.003) forms spontaneously from interactions with the environment, followed by a decreasing concentration of the spin species associated with photocatalytic activity (g = 2.004). Such behavior provides initial implications that a degradation process occurs over time.

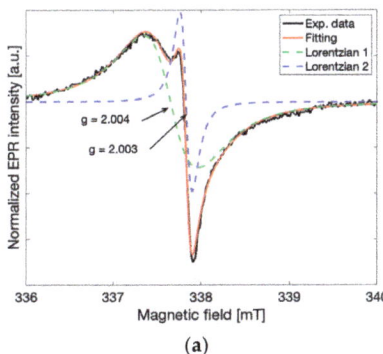
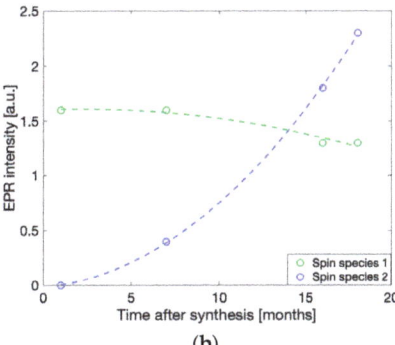

Figure 3. (**a**) The EPR spectrum of gCN (CN-M) shown with individual Lorentzian fits assuming a model comprised of the superposition of two Lorentzian lines, where each independent Lorentzian line is shown at g = 2.004 ± 0.00003 (*green*) and g = 2.003 ± 0.00003 (*blue*). The overall fit is shown as a sum of both individual Lorentzian components (*red*). (**b**) The time-dependent change in the observed EPR peak-to-peak intensities of the two spin species estimated by the fitting procedure is shown in (**a**). The color-scheme corresponding to the individual Lorentzian components is preserved throughout.

In newly synthesized gCN materials, the EPR spectral component corresponding to spin species 2 is negligible, so the gCN spectrum in principle may be described with a single Lorentzian lineshape and has been shown by previous groups where a single symmetric Lorentzian lineshape in the gCN EPR spectrum was observed [2,4,5,8]. Deviation from a Lorentzian shape as presented in this work is only observable following significantly long times after synthesis. Nevertheless, evidence of the presence of the second spin species long after the synthesis has made it possible to reconstruct the contributions of both spin species in recently synthesized samples allowing for investigation of the potential effects on photocatalytic activity on a relevant timescale and are herein reported.

2.2. Saturation and Relaxation Behavior

To investigate the independent behavior of the two observed Lorentzian contributions to the gCN EPR spectrum, power saturation behavior was recorded and the fitting procedure described previously was applied to each EPR spectrum recorded at varying microwave power levels. Power saturation curves interrogate the dependence of EPR amplitude on the applied B_1 field such that saturation behavior may be used for the determination of electronic longitudinal (T_1 or spin-lattice) and transverse (T_2 or spin-spin) relaxation rates of the system because these processes are closely correlated allowing for their estimation from CW-EPR experiments [10,11].

EPR spectra were recorded sequentially using microwave powers from 0.1 mW to 100 mW while keeping all other parameters constant. The maximum integrated intensities of the recorded spectra were plotted against B_1, which is proportional to the square root of microwave power (Figure 4, *black circles*). Because the spectra within this power range may be fit using two Lorentzian lines, the individual intensities of each of the two lines were constructed via the previously described fitting routine and were similarly plotted against B_1 (Figure 4, g = 2.004, *green circles*, and g = 2.003, *blue circles*), to yield simulated power saturation curves. Fits to the power saturation data were performed using the analytical formula for homogeneous saturation as follows,

$$I = \frac{B_1 I_m^0}{1 + B_1^2 \gamma^2 T_1 T_2} \qquad (2)$$

where I represents the integrated intensity of the EPR signal, B_1 is the magnetic field of the microwaves, $I_m^0 = \lim_{B_1 \to 0} \left(\frac{I}{B_1}\right)$, γ is the gyromagnetic ratio, T_1 is the spin-lattice relaxation time, and T_2 is the spin-spin relaxation time [11]. This model provides an estimation of T_1 relaxation times under the assumption that T_2 may be determined by the linewidth of a purely Lorentzian signal and no additional unresolved contributions to relaxation are present [11]. Both simulated saturation curves constructed based on individual signal intensities were found to be adequately described independently using a homogeneous saturation model (Figure 4, *green and blue dashed lines*) allowing for the conclusion that the gCN EPR spectrum consists of two spin centers at approximately g = 2.004 and g = 2.003. Since the involved spin species are not identical, the overall saturation of the system cannot be described by a homogeneous saturation model. Due to the complexity of non-homogeneous saturation and the many possible processes involved, an analytical description for the general case is not valid; however, the saturation behavior of gCN resulting from the recorded EPR spectra may be described as the sum of two separate power saturation curves within reasonable error (Figure 4, *red dashed line*). The significantly higher experimental values when compared to those obtained from the sum of the independently derived saturation curves could be explained by consideration of the integrated noise contributions to the overall intensity, which is not considered in the fitting routine [12]. Such discrepancies between the observed and calculated intensities may also arise from the assumption of absolute independence with respect to the simulations of the two spin systems, which is unlikely given their close proximity to one another. Therefore, additional weakly coupled interactions resulting in cross-correlations between the spin centers might occur and affect the saturation and relaxation behavior of each individual spin system [13,14]. Nevertheless, relaxation behavior may be qualitatively determined from these experiments such that spin species 1 (g = 2.004) which is represented by a wider EPR linewidth and saturation at lower microwave powers would typically correspond to faster T_2 and slower T_1, respectively, in comparison to spin species 2 (g = 2.003), which is instead described by a narrower EPR linewidth and later saturation, which would conversely have slower T_2 and faster T_1, respectively. This routine provides only an estimation and is not a direct observation; therefore, these results were further evaluated using complimentary pulse EPR and continuous wave saturation recovery EPR experiments.

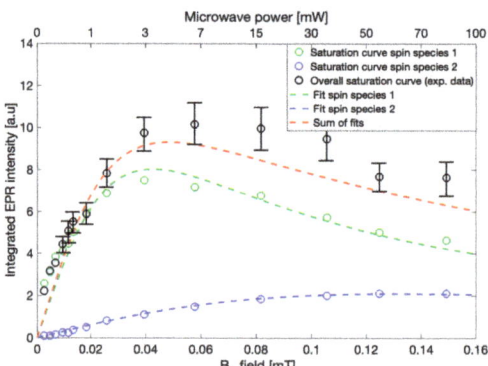

Figure 4. Simulated power saturation curves recorded via CW-EPR for the independent spin species identified in Figure 3a shown with the color scheme preserved (g = 2.004, *green*, g = 2.003, *blue*) with fits to the simulated data shown (*green and blue dashed lines*) assuming homogeneous saturation behavior. The experimentally recorded power saturation data are shown (*black open circles*) with the sum of the two simulated power saturation curves overlaid (*red dashed line*).

Following qualitative analysis of relaxation via CW experiments, relaxation time constants of gCN materials were measured directly by CW saturation recovery EPR techniques at X-band and by pulse EPR techniques at Q-band. CW saturation recovery measurements were preferentially performed at X-band due to the presumably very short T_2 time constants which therefore render pulse techniques at this frequency ineffective due to low echo intensities after consideration of the dead time of the instrument. The T_1 times estimated from the constructed saturation curves agree qualitatively with those obtained via continuous wave saturation recovery experiments. The relaxation times observed in gCN materials are not reported in the literature; however, these values may be compared with those observed in similar materials, for example, N-doped graphene oxide is reported to have similarly long T_1 relaxation [15–17]. Because only T_1 relaxation times increase with microwave frequency and T_2 is primarily frequency-independent [18,19], the two-pulse Hahn-echo sequence for measuring T_2 relaxation at Q-band was used for T_2 relaxation measurements. The resulting echo lasted sufficiently beyond the dead time of the instrument to enable not only T_2 measurements but also the three-pulse inversion recovery sequence for measuring T_1 relaxation at Q-band. In all experiments for all gCN compounds, similar and consistent results were obtained. For all relaxation data, fits using a biexponential function resulted in lower standard deviation in a residual analysis (Figures S1–S3 from Supplementary Materials). Mono-exponential functions were considered, but the residual analysis demonstrated nonlinear deviation above the noise levels in both cases. The spin–lattice (T_1) relaxation times and spin–spin (T_2) relaxation times varied by factors of only about two to three between all samples investigated and may be explained by differences in the morphologies of the samples due to diverse synthesis pathways [5], different spin concentrations influencing dipole coupling, and varying magnitudes of contributions from the first and second species and the resulting effects of cross correlated relaxation [13,14]. Further descriptions of the relaxation data and the various fitting approaches may be found in the supporting information. The multi-component relaxation behavior observed via CW saturation recovery at X-band and via pulse experiments at Q-band provide further support that two spin species are present in the gCN systems investigated.

A summary of relaxation times observed is given in Table 1. Longer T_1 values, which correspond to a lower degree of spin-lattice interactions, together with shorter T_2 values, which correspond to a greater contribution from spin-spin interactions, may indicate high delocalization and give some idea of the mobile character of spin species 1 (g = 2.004). These spins are primarily localized to the π-conjugated structure of the triazine rings and are responsible for photocatalytic activity [4]. On the contrary, shorter T_1 values and longer T_2 values observed for spin species 2 (g = 2.003) indicate a much greater interaction with the lattice or surrounding environment and a lower contribution from spin–spin interactions which may instead be correlated with an immobilized spin system and suggests that this species may not directly participate in the photocatalytic activity of the material.

Table 1. Relaxation time constants observed in continuous wave saturation recovery EPR experiments at X-band, two-pulse echo decay EPR experiments at Q-band, and three-pulse inversion recovery EPR measurements at Q-band after fitting with a biexponential model.

	T_1 [µs] X-Band CW Saturation Recovery		T_2 [ns] Q-Band Hahn-Echo		T_1 [µs] Q-Band Inversion Recovery	
	1st Species	2nd Species	1st Species	2nd Species	1st Species	2nd Species
CN-M	40 ± 10	4 ± 1	590 ± 30	3900 ± 200	450 ± 90	90 ± 20
CN-U	70 ± 10	4 ± 1	1300 ± 300	4800 ± 1200	370 ± 70	80 ± 10
CN-TU	-	-	700 ± 100	2600 ± 400	410 ± 60	100 ± 20
CN-CM	100 ± 20	15 ± 3	500 ± 100	1700 ± 400	700 ± 100	130 ± 20
CN-CMB	50 ± 5	2.0 ± 0.2	700 ± 50	1570 ± 80	520 ± 30	85 ± 5

2.3. Light Response

To understand the EPR signal response to light and therefore define the probabilities of each spin species exerting any effects on gCN photocatalytic performance, light-dependent EPR measurements were performed at X-band. For all gCN materials investigated, an increase in the resulting EPR signal was observed in response to irradiation by visible light. The EPR spectra recorded under dark conditions and under continuous light irradiation for the melamine-based gCN material are shown in Figure 5a. After applying the fitting routine described previously to differentiate the contributions from each of the spin species observed (Figure 5b), the EPR spectrum under light irradiation shows that the low field, wider component corresponding to spin species 1 ($g = 2.004$) increases by 60% when under irradiation, while the high field, narrower component corresponding to spin species 2 ($g = 2.003$) remains unchanged. Similar light-response behavior was observed for all investigated gCN materials and wavelengths of light in the range 400–700 nm. This demonstrates that only spin species 1 is light-active while species 2 which appears months after synthesis is light-inactive. Prolonged irradiation using white light resulted in further increase in the EPR peak-to-peak intensity of spin species 1 while the EPR intensity of spin species 2 remained unchanged. Similarly, the reverse effect was observed when removing the sample from light. After keeping the gCN materials in the dark for prolonged times, decreases in the EPR intensity of spin species 1 were observed while the EPR intensity of spin species 2 remained unchanged. These findings bring forth the implication that spin species 1 is likely associated with the spins responsible for photocatalytic activity [2–4]. The light-inactive character of spin species 2 instead implies that these spins likely do not contribute positively to photocatalysis. Rather, in conjunction with the time-dependent increase in the EPR intensity of spin species 2 following synthesis of the material, it is likely that this species may be indicative of the formation of degradation products within the gCN material. In much later stages of this study, a third spin species was detected in some but not all of the gCN samples investigated; however, this contribution to the overall signal was significantly lower than the two signals reported and was not well reproduced such that it was not considered for the further analysis (Figure S4 from Supplementary Materials). This signal likely corresponds to further degradation processes and may be attributed to an additional localized spin species as it does not respond to light; however, a thorough analysis of this contribution was beyond the scope of this work.

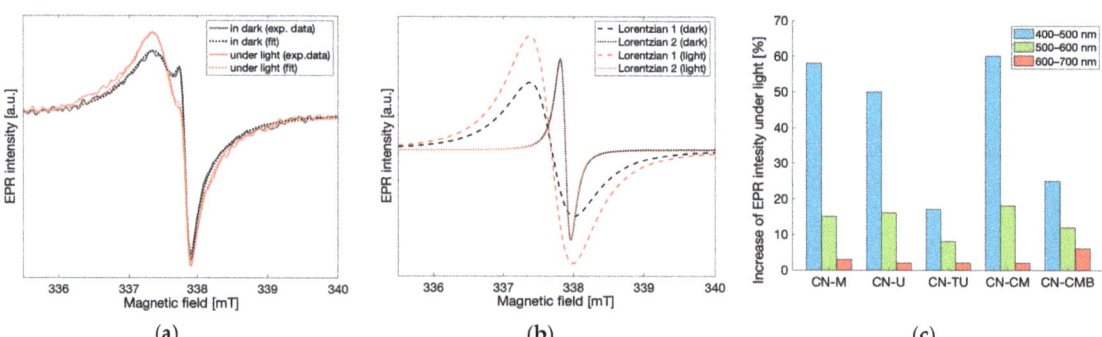

Figure 5. (a) EPR spectra of gCN (CN-M) recorded under dark conditions and under light irradiation (λ = 500–600 nm). (b) Individual Lorentzian lines obtained from the fits of the light irradiated and dark state spectra (c) The relative change of the EPR peak-to-peak intensity of spin species 1 ($g = 2.004$) under light irradiation using varying wavelengths of light compared to the EPR intensity recorded in the dark state for different gCN samples.

To better investigate the light-dependence of the EPR signals in the spectra obtained for gCN, EPR measurements were performed for all gCN samples while exposed to continuous light irradiation using the following 100 nm bands: 400–500 nm, 500–600 nm, 600–700 nm and in the dark state as a reference. Each recorded spectrum was fit using two Lorentzian lines, as described previously, and the relative change of the EPR peak-to-peak intensities between the dark state and under illumination at each of the irradiation bands for both Lorentzian lines was calculated independently. The relative changes in the EPR intensity at different bandwidths of excitation were normalized to the same photon flux while the linear dependence of the light-induced increase in EPR intensity with respect to photon flux was verified separately using a halogen cold light source (Schott KL 2500 LCD and Thurlby Thandar TSX1820P). As stated previously, the intensity of spin species 2 remained unchanged in all irradiation conditions. Therefore, only the relative change in the EPR intensity of spin species 1 is shown in Figure 5c. For all gCN materials investigated, the maximum increase in EPR intensity was observed when the spectrum was recorded using light irradiation bandwidths of 400–500 nm, while the increase in EPR intensity was significantly lower for the 500–600 nm bandwidth and even smaller for the 600–700 nm bandwidth. Noticeably, this behavior correlates very well with UV-Vis spectra previously recorded for gCN materials [20,21]. Though the EPR spectrum of gCN materials is typically interpreted as a single component, separating gCN EPR spectra into multiple components and considering only the light-active species in the assessment of light-dependent properties leads to better predictions of photocatalytic activity [4,7]. While this experiment allows for the comparison of the light response within the material under irradiation from different wavelengths, comparability between the overall activity of different gCN materials with respect to surface effects must be considered. It has been shown that photocatalytic activity mainly occurs on the surface of the material [22–24]. However, due to morphological differences within the materials impacting the diffraction of the incident light, as well as variations in porosity and therefore variations in both density and surface area, the number of spins exposed to irradiation may vary greatly between different gCN materials. Therefore, a comparison between materials according to the light-induced EPR response was not considered due to the difference in the responses between samples (Figure 5c) which primarily corresponds to the difference in numbers of spins accessible by light irradiation and not directly the total number of photoactive spins within the material.

2.4. Oxidation Effects

In order to investigate the origin of spin species 2 ($g = 2.003$) which was thought to result from interactions with the environment leading to the observed time-dependent changes in the EPR spectra, attempts were made to elicit similar processes artificially while recording the EPR spectrum both before and after the elicited process. The gCN materials were stored in non-sealed vials open to the air, where both reactive oxygen and water are present [25]. Since gCN is a photocatalyst, it is possible that H_2 may be formed via water splitting reactions using water from the air even without special conditions. To determine if this is the case, the relative humidity was increased in the sample tube and any resulting effects on the observed EPR lineshape were recorded. This was performed in two ways: by blowing water vapor into the sample tube and by adding water dropwise directly to the sample tube. In neither of these experiments were any changes to the lineshape observed.

To test the effects of oxygen in the environment on the spectral lineshape, the sample tube with gCN powder was exposed to an oxygen-enriched environment by replacing the atmosphere of the EPR sample tube with oxygen. EPR spectra were recorded immediately after oxygen was introduced and continuously for 8 h. In Figure 6a spectra taken in the oxygen-enriched environment and in normal atmospheric conditions are shown for comparison. Both an increase in the signal corresponding to spin species 2 and a decrease in the signal corresponding to spin species 1 were observed. In Figure 6b Lorentzian lines calculated from the fitting routine described above and corresponding to spin species 1 ($g = 2.004$, *dotted lines*) and spin species 2 ($g = 2.003$, *dashed lines*) are shown under normal atmospheric conditions (*black*) and oxygen-enriched conditions (*red*). The observed relative change in the peak-to-peak intensities of the Lorentzian lines demonstrates that concentrations of spin species 2 increased while concentrations of spin species 1 decreased after oxygen enrichment in a manner similar to those observed to be taking place slowly over time. Direct oxidation of the gCN material was attempted via treatment with aqueous H_2O_2 solution (3%) and the EPR response before and after treatment was recorded. The obtained results demonstrated a similar effect where the signal from spin species 2 increases in intensity while the signal from spin species 1 decreases. This effect is quickly followed by additional degradation of the material, characterized by numerous additional overlapping signals (Figure S5 from Supplementary Materials) [9]. Further analysis of these results was not pursued due to low SNR from the increased microwave absorption of aqueous solutions. From these experiments taken together, it may be concluded that a local increase in oxygen concentration interacting with the gCN material leads to similar spectral changes to those observed in the time-dependent response, providing unambiguous support for oxidation processes as the primary influence on the time-dependent appearance of spin species 2 and may be attributed to new spin centers formed via the interaction of gCN with atmospheric oxygen.

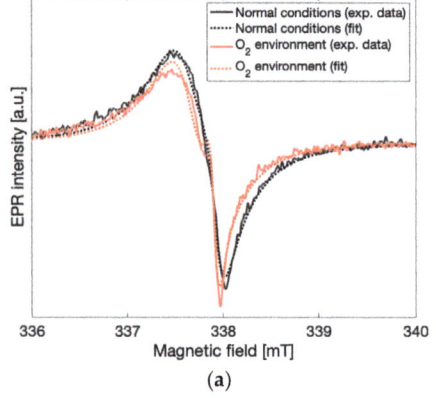 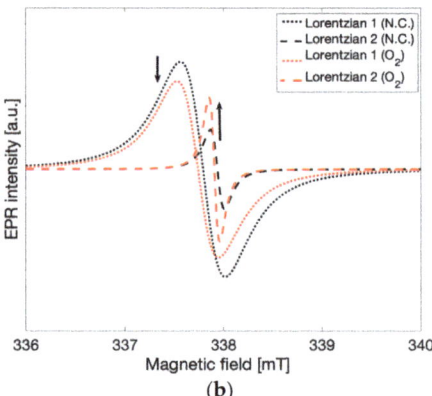

Figure 6. (a) The change in the EPR spectrum observed under normal atmospheric conditions and after exposure to an oxygen enriched environment. (b) The individual Lorentzian lines obtained from the fitting routine correspond to spin species 1 (g = 2.004) and spin species 2 (g = 2.003) for both atmospheric and oxygen enriched conditions, shown on the same scale as the experimental data.

2.5. Impact on Catalytic Performance

To determine the effects from the time-dependent structural changes in gCN materials observed by EPR on their photocatalytic performance, hydrogen evolution measurements on fresh gCN materials and gCN materials synthesized 20 months prior were conducted. Among the gCN materials studied, CN-CM and CN-CMB were chosen on account of their notably high reproducibility and stability in terms of hydrogen evolution rates [21]. As shown in Figure 7, both CN-CM and CN-CMB materials demonstrated a decrease in the observed H_2 evolution rate over time, with the appearance of spin species 2 in the EPR spectrum. The introduction of spin species 2 is accompanied by a reduction in spin species 1 which is presumed to be the spin species predominantly responsible for photocatalytic activity and correlates well with a decrease in the overall concentration of photoactive electrons over time, negatively impacting the photocatalytic H_2 evolution rate. The oxidative processes in the gCN materials are detrimental for the photocatalytic performance of both CN-CM and CN-CMB and is representative of all gCN materials investigated.

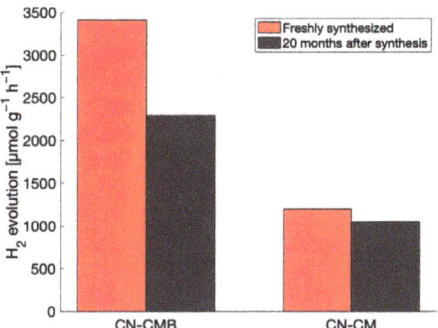

Figure 7. Hydrogen evolution rates observed for two different gCN materials, CN-CM and CN-CMB, when measured immediately after synthesis and 20 months after synthesis.

3. Discussion

In Figure 8, a structural model is given depicting the potential sites for spin centers that would likely result in the EPR signal (spin species 2, g = 2.003) observed to appear and increase over time in gCN materials. While spin species 1 (g = 2.004) is attributed to electrons delocalized over the entire π-conjugated structure of the triazine unit, it is now suggested that the signal associated with spin species 2 originates from oxidation-produced radicals localized at carbon atoms [26]. These radicals may be formed by either the introduction of an oxygen-centered species (denoted with R in Figure 8) or via oxidation of carbon atoms directly, most likely at the border of the triazine unit, without incorporation of the reactive oxygen into the molecule [17,25,27]. Alternatively, reactive oxygen could attack the nitrogen atoms to form nitrogen-centered radicals; however, the g-value of the narrow line (spin species 2, g = 2.003) is in much better agreement with the formation of carbon-centered radicals [28]. Formation of radicals via oxidative damage destroys the local electronic configuration in a way that decreases the total area of delocalization within the π-conjugated structure accessible by the electrons [6,25]. Thus, with increasing concentration of radicals formed from oxidative damage in gCN materials, the concentration of photoactive delocalized electrons decreases simultaneously. This results in quenching of the overall photocatalytic activity.

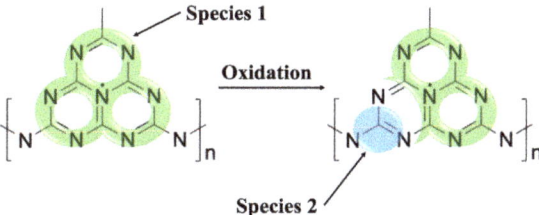

Figure 8. Potential oxidative processes in the gCN structure where delocalization of spin species 1 (g = 2.004) is highlighted in green while the introduction of spin species 2 (g = 2.003) associated with radicals formed via oxidation is highlighted in blue.

The oxidation mechanism proposed aligns well with the herein reported observations based on the experiments performed. EPR spectra recorded from CW experiments reveal simultaneous contributions from multiple spin centers as early as six months after synthesis of gCN materials, where two spin species, one of which is highly mobile and likely responsible for photocatalytic activity as shown by the observed relaxation behavior, light-response and literature support [2,4,7,8], while another demonstrates both relaxation behavior and a lack of any light-response typical of immobilized states or isolated radicals [17,27]. Both processes, the increasing concentration of spin species 2 (g = 2.003) and the decreasing concentration of spin species 1 (g = 2.004) were observed to occur with time. Evidence of the influence of reactive oxygen species, and not water present in the atmosphere, on the EPR lineshape and thus the spin centers present is provided by the EPR measurements performed both before and after exposure to water, oxygen-enriched environments, and H_2O_2 solutions. The detrimental behavior of these interactions is demonstrated by the reduced photocatalytic H_2 production rates in the newly synthesized and oxidized gCN materials investigated.

These findings concerning the long-term stability of gCN materials indicate a need for further investigation of the degradation processes and how they might be alleviated in large-scale applications. As has been demonstrated by the observed reductions in photocatalytic performance, possibilities to prevent oxidative damage are necessary enroute to functional devices. Methods of oxidation may be reduced by close attention to handling of the materials during synthesis and while in use such that the H_2 production rate is maintained over long durations.

4. Materials and Methods

4.1. Synthesis

The gCN samples investigated in the present work were synthesized from different precursors (Figure 1a) with slightly different synthesis procedures.

4.1.1. Synthesis of Supramolecular Assemblies

Cyanuric acid-melamine (CM) complex was prepared by mixing cyanuric acid (0.51 g, 4 mmol) and melamine (0.50 g, 4 mmol) in a 1:1 molar ratio in 40 mL DI water for 12 h in an automatic shaker (KS-260, IKA-Werke GmbH & Co. KG, Staufen, Germany). The obtained solid precursor was filtered and then dried at 60 °C in a vacuum oven. Cyanuric acid-melamine-barbituric acid (CMB) complex was synthesized by mixing in a 1:1:0.05 molar ratio, following the same procedure as for the CM complex.

4.1.2. Preparation of CN Materials

The gCN materials were prepared via thermal condensation of each precursor; melamine (M), urea (U), thiourea (TU), cyanuric acid and melamine following supramolecular preorganization methods (CM), and cyanuric acid, barbituric acid, and melamine following supramolecular preorganization methods (CMB) at 550 °C for 4 h under an inert N_2 atmosphere in a muffle furnace (heating rate of 2.5 °C min^{-1} from room temperature). All gCN products were collected and labelled as CN-X, where X represents the corresponding precursor.

4.2. Electron Paramagnetic Resonance (EPR) Measurements

Powder gCN samples were placed in quartz/suprasil 4 mm O.D. EPR tubes. Since the density and thus the spin concentrations vary considerably between different gCN powders, from 10^{12} spins per milligram for CN-M to 10^{14} spins per mg for CN-CMB, the volume of each sample was carefully controlled rather than the mass. Each tube was filled with approximately 0.25 cm^3 powder, which is equal to 2 cm sample height in the EPR tube and is equivalent to the active space of the resonant cavity. EPR measurements were performed using a Magnetech MS-5000 benchtop X-band spectrometer with an operating frequency of 9.45 GHz at room temperature and ambient light access. Modulation amplitude and frequency were 0.1 mT and 100 kHz. CW spectra were recorded using a microwave power of 15 mW, while for power saturation measurements, microwave power was varied from 0.05 mW to 100 mW. Comparative CW spectra were normalized to the integrated intensity of the absorption signal, when appropriate. The standard deviation and total variance of the g-values reported were calculated from three consecutive CW measurements using N@C$_{60}$ as a reference standard and were found to be ± 0.00003 and ± 0.00006, respectively.

Pulse EPR measurement were performed using an Elexsys E580 spectrometer (Bruker Biospin, Ettlingin, Germany) with a Q-band resonator (33.75 GHz) and microwave power of 20 mW. Samples were placed in quartz/suprasil 1.6 mm EPR tubes. Saturation recovery measurements were performed using a modified Bruker E500T spectrometer described previously [29].

For light-dependent measurements a Magnettech MS-5000 EPR spectrometer and a supercontinuum laser, a SuperK Fianium equipped with a Varia filter (NKT Photonics, Birkerød, Denmark), were used. The pulse repetition rate of the laser was 78.2 MHz, much faster than the rate of the CW EPR measurements. The EPR spectrometer and laser were connected in such a manner that the laser beam was directed into the middle of the resonator and uniformly illuminated one face of the sample volume, while outside of the resonator dark conditions were maintained.

EPR spectra were recorded for each gCN material investigated. CW spectra presented herein are shown for CN-M, but the spectral descriptions and resulting observations and conclusions are relevant for all gCN materials investigated unless otherwise stated.

4.3. Photocatalytic H_2 Production Measurements

The hydrogen evolution reaction was conducted under a constant temperature of 25 °C in a sealed quartz vessel held by a jacketed beaker thermally regulated with a circulating cooling system. In each, 15 mg of CN material was suspended via sonication for 15 min in an aqueous solution containing 17.1 mL of DI water, 1.9 mL of TEOA as the hole scavenger, and 19.6 µL of H_2PtCl_6 solution (3%wt Pt relative to CN-X photocatalyst mass). After purging with Ar to remove residual air in the vessel, Pt cocatalyst nanoparticles were in-situ photodeposited on the CN-X catalyst surface under white light-emitting diode (LED) light irradiation (BXRA-50C5300, 100 W, λ > 410 nm, Bridgelux, Fremont, CA, USA) for 0.5 h with continuous stirring (600 rpm). Gas chromatography (Agilent 7820 GC) was employed to monitor the amount of evolved H_2 gas in the headspace every 0.5 h.

Supplementary Materials: The following supporting information can be downloaded at: https://www.mdpi.com/article/10.3390/molecules28186475/s1, Figure S1: T_2 relaxation decay curve measured via the Hahn-echo sequence in Q band and fit with a biexponential function for CN-CMB. Figure S2: T_1 relaxation recovery curve measured via the 3-pulse inversion recovery sequence in Q band and fit with a biexponential function for CN-CMB. Figure S3: Example of residual analysis after subtraction of fits via either exponential or biexponential functions for T_1 relaxation data. Figure S4: EPR spectrum of gCN fit with a superposition of 3 Lorentzian lines. Figure S5: EPR spectrum of gCN after adding H_2O_2.

Author Contributions: E.K., E.S. and J.E.M. defined the goals of the research and designed the experiments. E.K., E.S. and J.E.M. performed all EPR experiments, data processing and EPR simulations. A.A., N.K. and M.S. prepared the samples and performed hydrogen evolution experiments. E.K., J.E.M. and R.W.M. performed light-irradiated EPR experiments. E.K., E.S., K.L. and J.E.M. evaluated the results of the experiments. E.K. and J.E.M. wrote the manuscript. The article was revised by all authors. All authors have read and agreed to the published version of the manuscript.

Funding: This research has been supported by the Bundesministerium für Bildung und Forschung (EPRoC grant no. 01186916/1). This research was carried out within the Helmholtz International Research School (HIRS-0008), "Hybrid Integrated Systems for Conversion of Solar Energy" (HI-SCORE), an initiative co-funded by the "Initiative and Networking Fund of the Helmholtz Association".

Data Availability Statement: Data will be made available upon reasonable request.

Acknowledgments: We acknowledge the highly beneficial discussions with Silvio Künstner, Gianluca Marcozzi, Michele Segantini and Boris Naydenov. We also acknowledge Gareth and Sandra Eaton for their advice, experimental expertise, and use of their facilities for saturation recovery and pulse experiments.

Conflicts of Interest: The authors declare no conflict of interest.

Sample Availability: Not applicable.

References

1. Liang, J.; Jiang, Z.; Wong, P.K.; Lee, C.-S. Recent Progress on Carbon Nitride and Its Hybrid Photocatalysts for CO_2 Reduction. *Sol. RRL* **2021**, *5*, 2000478. [CrossRef]
2. Ong, W.-J.; Tan, L.-L.; Ng, Y.H.; Yong, S.-T.; Chai, S.-P. Graphitic Carbon Nitride (g-C_3N_4)-Based Photocatalysts for Artificial Photosynthesis and Environmental Remediation: Are We a Step Closer To Achieving Sustainability? *Chem. Rev.* **2016**, *116*, 7159–7329. [CrossRef] [PubMed]
3. Wang, X.; Maeda, K.; Thomas, A.; Takanabe, K.; Xin, G.; Carlsson, J.M.; Domen, K.; Antonietti, M. A Metal-Free Polymeric Photocatalyst for Hydrogen Production from Water under Visible Light. *Nat. Mater.* **2009**, *8*, 76–80. [CrossRef] [PubMed]
4. Xia, P.; Cheng, B.; Jiang, J.; Tang, H. Localized π-Conjugated Structure and EPR Investigation of g-C3N4 Photocatalyst. *Appl. Surf. Sci.* **2019**, *487*, 335–342. [CrossRef]
5. Barrio, J.; Shalom, M. Rational Design of Carbon Nitride Materials by Supramolecular Preorganization of Monomers. *ChemCatChem* **2018**, *10*, 5573–5586. [CrossRef]
6. Thomas, A.; Fischer, A.; Goettmann, F.; Antonietti, M.; Müller, J.O.; Schlögl, R.; Carlsson, J.M. Graphitic Carbon Nitride Materials: Variation of Structure and Morphology and Their Use as Metal-Free Catalysts. *J. Mater. Chem.* **2008**, *18*, 4893–4908. [CrossRef]
7. Actis, A.; Melchionna, M.; Filippini, G.; Fornasiero, P.; Prato, M.; Salvadori, E.; Chiesa, M. Morphology and Light-Dependent Spatial Distribution of Spin Defects in Carbon Nitride. *Angew. Chem. Int. Ed.* **2022**, *61*, e202210640. [CrossRef]
8. Tu, W.; Xu, Y.; Wang, J.; Zhang, B.; Zhou, T.; Yin, S.; Wu, S.; Li, C.; Huang, Y.; Zhou, Y.; et al. Investigating the Role of Tunable Nitrogen Vacancies in Graphitic Carbon Nitride Nanosheets for Efficient Visible-Light-Driven H_2 Evolution and CO_2 Reduction. *ACS Sustain. Chem. Eng.* **2017**, *5*, 7260–7268. [CrossRef]
9. Dvoranová, D.; Barbieriková, Z.; Mazúr, M.; García-López, E.I.; Marcì, G.; Lušpai, K.; Brezová, V. EPR Investigations of Polymeric and H_2O_2-Modified C_3N_4-Based Photocatalysts. *J. Photochem. Photobiol. A Chem.* **2019**, *375*, 100–113. [CrossRef]
10. Eaton, G.R.; Eaton, S.S.; Barr, D.P.; Weber, R.T. *Quantitative EPR*; Springer Vienna: Vienna, Austria, 2010; ISBN 978-3-211-92947-6.
11. Poole, C. *Electron-Spin-Resonance*, 2nd ed.; A Comprehensive Treatise on Experimental Techniques; CRC Press: Boca Raton, FL, USA, 1983; ISBN 0471046787.
12. Ruppert, M.G.; Bartlett, N.J.; Yong, Y.K.; Fleming, A.J. Amplitude Noise Spectrum of a Lock-in Amplifier: Application to Microcantilever Noise Measurements. *Sens. Actuators A Phys.* **2020**, *312*, 112092. [CrossRef]
13. Ran, J.; Ma, T.Y.; Gao, G.; Du, X.W.; Qiao, S.Z. Porous P-Doped Graphitic Carbon Nitride Nanosheets for Synergistically Enhanced Visible-Light Photocatalytic H2 Production. *Energy Environ. Sci.* **2015**, *8*, 3708–3717. [CrossRef]
14. Groenewolt, M.; Antonietti, M. Synthesis of G-C_3N_4 Nanoparticles in Mesoporous Silica Host Matrices. *Adv. Mater.* **2005**, *17*, 1789–1792. [CrossRef]
15. Tampieri, F.; Tommasini, M.; Agnoli, S.; Favaro, M.; Barbon, A. N-Doped Graphene Oxide Nanoparticles Studied by EPR. *Appl. Magn. Reson.* **2020**, *51*, 1481–1495. [CrossRef]
16. Tadyszak, K.; Chybczyńska, K.; Ławniczak, P.; Zalewska, A.; Cieniek, B.; Gonet, M.; Murias, M. Magnetic and Electric Properties of Partially Reduced Graphene Oxide Aerogels. *J. Magn. Magn. Mater.* **2019**, *492*, 165656. [CrossRef]
17. Fanchini, G.; Ray, S.C.; Tagliaferro, A.; Laurenti, E. Paramagnetic Centres and Microstructure of Reactively Sputtered Amorphous Carbon Nitride Thin Films. *Diam. Relat. Mater.* **2002**, *11*, 1143–1148. [CrossRef]
18. Robinson, B.H.; Haas, D.A.; Mailer, C. Molecular Dynamics in Liquids: Spin-Lattice Relaxation of Nitroxide Spin Labels. *Science (1979)* **1994**, *263*, 490–493. [CrossRef]
19. Eaton, S.S.; Eaton, G.R. Relaxation Mechanisms. *eMagRes* **2016**, *5*, 1543–1556. [CrossRef]
20. Chen, Y.; Wang, B.; Lin, S.; Zhang, Y.; Wang, X. Activation of N- * Transitions in Two-Dimensional Conjugated Polymers for Visible Light Photocatalysis. *J. Phys. Chem. C* **2014**, *118*, 29981–29989. [CrossRef]
21. Shalom, M.; Inal, S.; Fettkenhauer, C.; Neher, D.; Antonietti, M. Improving Carbon Nitride Photocatalysis by Supramolecular Preorganization of Monomers. *J. Am. Chem. Soc.* **2013**, *135*, 7118–7121. [CrossRef]
22. Zheng, Q.; Durkin, D.P.; Elenewski, J.E.; Sun, Y.; Banek, N.A.; Hua, L.; Chen, H.; Wagner, M.J.; Zhang, W.; Shuai, D. Visible-Light-Responsive Graphitic Carbon Nitride: Rational Design and Photocatalytic Applications for Water Treatment. *Environ. Sci. Technol.* **2016**, *50*, 12938–12948. [CrossRef]
23. Zheng, Y.; Lin, L.; Wang, B.; Wang, X. Graphitic Carbon Nitride Polymers toward Sustainable Photoredox Catalysis. *Angew. Chem.* **2015**, *127*, 13060–13077. [CrossRef]
24. Wang, S.; Zhang, J.; Li, B.; Sun, H.; Wang, S. Engineered Graphitic Carbon Nitride-Based Photocatalysts for Visible-Light-Driven Water Splitting: A Review. *Energy Fuels* **2021**, *35*, 6504–6526. [CrossRef]
25. Miller, D.J.; McKenzie, D.R. Electron Spin Resonance Study of Amorphous Hydrogenated Carbon Films. *Thin Solid. Film.* **1983**, *108*, 257–264. [CrossRef]
26. Silva, S.R.P.; Robertson, J.; Amaratunga, G.A.J.; Rafferty, B.; Brown, L.M.; Schwan, J.; Franceschini, D.F.; Mariotto, G. Nitrogen Modification of Hydrogenated Amorphous Carbon Films. *J. Appl. Phys.* **1997**, *81*, 2626–2634. [CrossRef]
27. Fanciulli, M.; Fusco, G.; Tagliaferro, A. Insight on the Microscopical Structure of A-C and a-C:H Thin Films through Electron Spin Resonance Analysis. *Diam. Relat. Mater.* **1997**, *6*, 725–729. [CrossRef]

28. Monclus, M.A.; Chowdhury, A.K.M.S.; Cameron, D.C.; Barklie, R.; Collins, M. The Effect of Nitrogen Partial Pressure on the Bonding in Sputtered CNx Films: Implications for Formation of β-C3N4. *Surf. Coat. Technol.* **2000**, *131*, 488–492. [CrossRef]
29. McPeak, J.E.; Quine, R.W.; Eaton, S.S.; Eaton, G.R. An X-Band Continuous Wave Saturation Recovery Electron Paramagnetic Resonance Spectrometer Based on an Arbitrary Waveform Generator. *Rev. Sci. Instrum.* **2019**, *90*, 24102. [CrossRef]

Disclaimer/Publisher's Note: The statements, opinions and data contained in all publications are solely those of the individual author(s) and contributor(s) and not of MDPI and/or the editor(s). MDPI and/or the editor(s) disclaim responsibility for any injury to people or property resulting from any ideas, methods, instructions or products referred to in the content.

Article

Zinc–Cobalt Oxide Thin Films: High Curie Temperature Studied by Electron Magnetic Resonance

Bogumił Cieniek [1,*], Ireneusz Stefaniuk [1], Ihor Virt [2,3], Roman V. Gamernyk [4] and Iwona Rogalska [1]

1 Institute of Materials Engineering, College of Natural Sciences, University of Rzeszow, Pigonia 1, 35-310 Rzeszow, Poland
2 Institute of Physics, College of Natural Sciences, University of Rzeszow, Pigonia 1, 35-310 Rzeszow, Poland
3 Institute of Physics, Mathematics, Economy and Innovation Technologies, Drohobych Ivan Franko State Pedagogical University, Stryiska 3, 82100 Drohobych, Ukraine
4 Department of Experimental Physics, Ivan Franko National University of Lviv, 79005 Lviv, Ukraine
* Correspondence: bcieniek@ur.edu.pl

Abstract: The material with a high Curie temperature of cobalt-doped zinc oxide embedded with silver-nanoparticle thin films was studied by electron magnetic resonance. The nanoparticles were synthesized by the homogeneous nucleation technique. Thin films were produced with the pulsed laser deposition method. The main aim of this work was to investigate the effect of Ag nanoparticles on the magnetic properties of the films. Simultaneously, the coexisting Ag^0 and Ag^{2+} centers in zinc oxide structures are shown. A discussion of the signal seen in the low field was conducted. To analyze the temperature dependence of the line parameters, the theory described by Becker was used. The implementation of silver nanoparticles causes a significant shift of the line, and the ferromagnetic properties occur in a wide temperature range with an estimated Curie temperature above 500 K.

Keywords: ferromagnetism at room temperature; electron magnetic resonance; low-field microwave absorption; zinc oxide; silver nanoparticles

Citation: Cieniek, B.; Stefaniuk, I.; Virt, I.; Gamernyk, R.V.; Rogalska, I. Zinc–Cobalt Oxide Thin Films: High Curie Temperature Studied by Electron Magnetic Resonance. *Molecules* **2022**, *27*, 8500. https://doi.org/10.3390/molecules27238500

Academic Editor: Yordanka Karakirova

Received: 28 October 2022
Accepted: 30 November 2022
Published: 2 December 2022

Publisher's Note: MDPI stays neutral with regard to jurisdictional claims in published maps and institutional affiliations.

Copyright: © 2022 by the authors. Licensee MDPI, Basel, Switzerland. This article is an open access article distributed under the terms and conditions of the Creative Commons Attribution (CC BY) license (https://creativecommons.org/licenses/by/4.0/).

1. Introduction

Zinc oxide (ZnO) is an inexpensive well-known semiconductor with potential in various applications [1–3], such as varistors [4] or sensors [5,6]. It is important that some of the zinc can be replaced by magnetic transition metal ions (TM) to create a metastable solid solution. Since both Zn^{2+} and Co^{2+} ions have nearly identical ion radii, doping ZnO with cobalt is most interesting [7]. Furthermore, parameters such as piezoelectricity and transparency in the visual region have attracted great interest from researchers in ZnO-based diluted magnetic semiconductors (DMS) [8,9] due to their possible technological applications in spintronics [10–13]. Using the modified Zener model, Dietl et al. suggested that p-type DMS based on ZnO could lead to a transition temperature greater than room temperature [14]. In this theory, p–d interactions are the cause of long-range magnetic coupling, but the studied ZnO samples are either insulating or conducting n-types. Some theoretical works using density functional theory (DFT) [15,16] show that n-type cobalt-doped ZnO shows ferromagnetism (FM) at room temperature (RTFM). Some research groups reported ferromagnetism in ZnO doped with transition metals with Curie temperatures (T_C) from 30 to 550 K [17–23], and some found antiferromagnetic, spin-glass, or paramagnetic behavior [9,24,25]. The existence of a ferromagnetic order in Co-doped ZnO is suggested to be attributed to double exchange [8] or the Ruderman–Kittel–Kasuya–Yosida (RKKY) interaction between Co ions [26]. Theoretical calculations show that ground-state ZnO with Co ions is spin-glass because of the short-range interactions between TM atoms [27].

ZnO doped with gold (Au) or silver (Ag) increases the photocatalytic activity of the composite by reducing electron-hole recombination and improving separation [28]. Silver nanoparticles (NP) have been investigated by many scientists because of their significant role in applications of visible light absorption [29,30]. Many works have described the synthesis nanocomposites of heterogeneous ZnO/Ag via a variety of synthetic routes for various applications, such as disinfection and wastewater treatment [31–39]. In addition, research confirms that the existence of Ag NPs on the ZnO surface reduces the intensity of electron magnetic resonance (EMR) signals and may lead to improved photodegradation efficiency [40].

Obtaining a homogeneous thin layer with *p*-type ZnO is a challenging task. One of the proposed methods is the addition of silver ions [41–43]. The pulsed laser deposition (PLD) method has a wide range of particle energies, allowing mainly Zn ions to penetrate deeper into the substrate, forming a mixed structure with the desired conductive type [43]. The main aim of this work was to investigate the influence of silver nanoparticles on ZnO doped with cobalt-ion thin film and to determine the changes in magnetic properties compared to that of layers without silver NPs.

2. Results and Discussion

Measurements of EMR were taken on samples with a quartz and silicon substrate. Angular dependence measurements of the $Zn_{0.8}Co_{0.2}O/Ag$ film on the quartz substrate were measured at room temperature; a summary is shown in Figure 1.

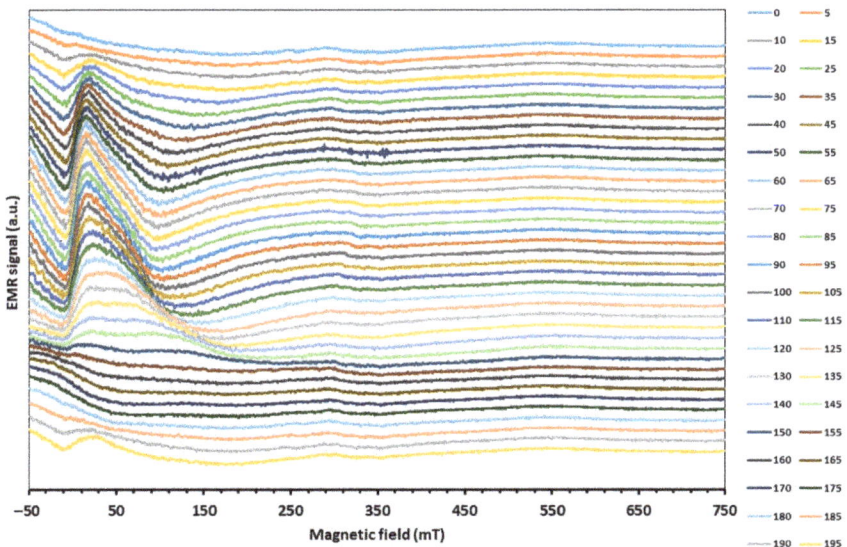

Figure 1. EMR spectra as a function of the angular dependence of $Zn_{0.8}Co_{0.2}O/Ag$ film.

The resulting angular dependence is characteristic of magnetic layers and the combination of magnetic and nonmagnetic layers. A strong anisotropy of the spectrum is observed, along with a change in the shape of the line, with the result that, at certain angles, the width of the line increases significantly as the intensity decreases, and the EMR line is no longer visible. However, a line visible all the time in the low field remains, the so-called low-field microwave absorption (LFMA), which is also called an indicator of FM properties for a large group of materials. The LFMA signal for ferrites and magnets is related to the beginning of the ordered phase and is a sensitive detector of magnetic ordering [44,45]. For soft magnetic materials, the signal is due to low-field processes of spin magnetization [46].

EMR measurements as a function of temperature were performed in two temperature ranges, from 300 to 500 K and from 97 to 300 K. Figure 2 shows the EMR spectra of $Zn_{0.8}Co_{0.2}O/Ag$ as a function of temperature in the range from 300 to 500 K. The measurement was performed at an angle when the EMR line was close to its extreme position, and it corresponds to an orientation of 115 degrees from the angular dependence in Figure 1.

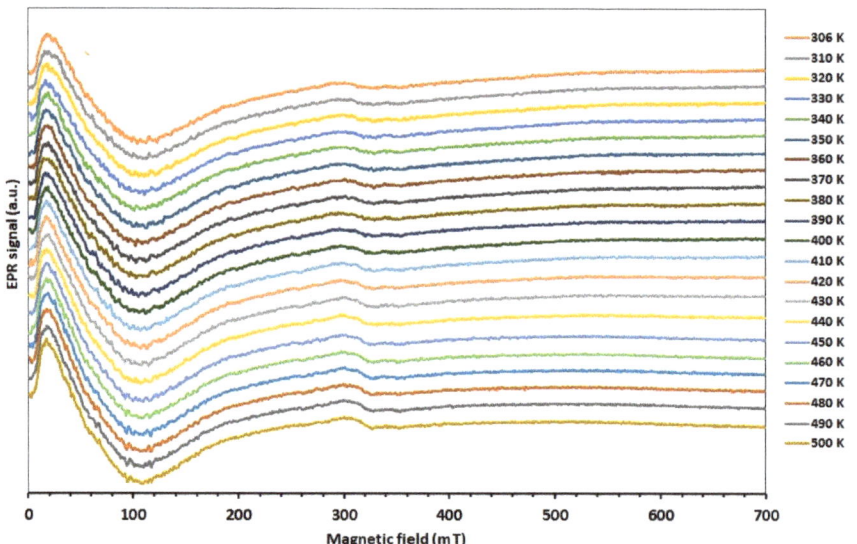

Figure 2. EMR spectra of $Zn_{0.8}Co_{0.2}O/Ag$ as a function of temperature in the range from 300 to 500 K.

In contrast to the angular dependence of the EMR spectrum, the temperature dependence over the entire temperature range studied does not show large changes in the shape and position of the EMR line. We can see a broad line moving in the direction of the low magnetic field for $Zn_{0.8}Co_{0.2}O/Ag$ compared to the sample without silver NPs (Figure 3) and to layers of Co-doped ZnO (in our previous papers [47–49]).

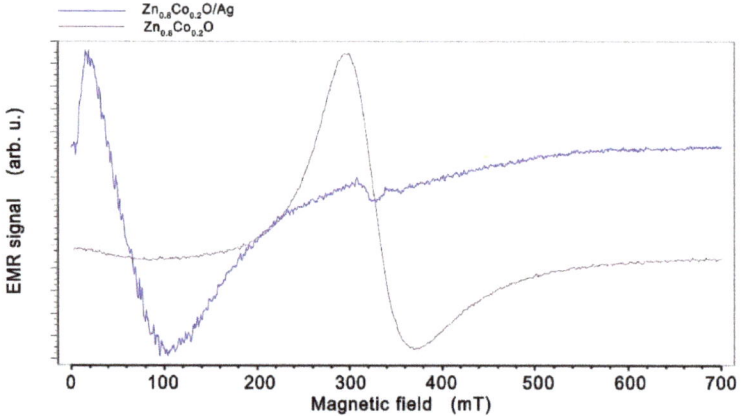

Figure 3. EMR spectra of $Zn_{0.8}Co_{0.2}O$ (purple) and $Zn_{0.8}Co_{0.2}O$ embedded with Ag NPs (blue), obtained at temperature 160 K.

Such a large shift of the EMR line can be explained by the appearance of ferromagnetic properties in $Zn_{0.8}Co_{0.2}O$ embedded with Ag NPs. A similar effect is observed in many works; for example, in spin-glasses or soft and hard magnetic layers [50–53]. The hard and soft layers of the spring magnets are coupled at the interfaces due to the strong exchange coupling between them. A high magnetic saturation is achieved by a soft magnet, whereas a high coercivity field is achieved by the magnetically hard material. The shift of the EMR line towards a low magnetic field for the layers of work [52] is related to the layer thickness and occurs when the thickness changes from 10 to 20 nm. In our sample, we observe conglomerates of nanoparticles with sizes in the order of 80 nm, as well as single nanoparticles, so, in addition to the EMR line, we also observe a spectrum from single silver nanoparticles. Low-intensity lines can also be seen in a field of about 340 mT. These were assigned to silver ions Ag^0 ($4d^{10}5s^1$) and/or Ag^{2+} and Ag^0 ($4d^9$) (Figure 4) (described in the literature [54]). This confirms that Ag^0 and Ag^{2+} centers can coexist simultaneously in zinc oxide structures, with Ag^+ being inactive in the EMR signal. Matching the EMR spectrum with the Dyson-type line results from a good fit to our sample $Zn_{0.8}Co_{0.2}O$ (Figure 4), and hence the line parameters were obtained: the peak-to-peak linewidth (H_{pp}), the EMR intensity (I), and the resonance field (H_r). Xepr software was used to analyze and determine the parameters of the EMR line; this is the standard EPR spectrometer software used to control and analyze the spectrum.

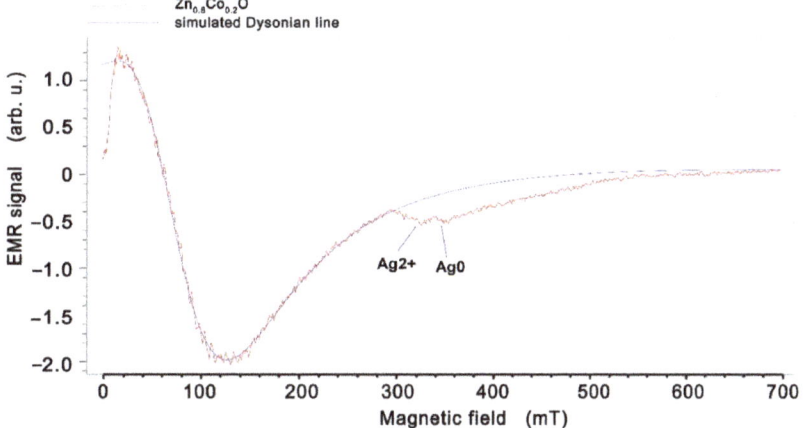

Figure 4. EMR spectra of $Zn_{0.8}Co_{0.2}O$ embedded with Ag NPs film obtained at temperature 300 K with the simulated Dyson-type line. Small intensity lines from Ag^{2+} and Ag^0 are described.

Figure 5 shows the intensities of the EMR line as a function of temperature in two temperature ranges.

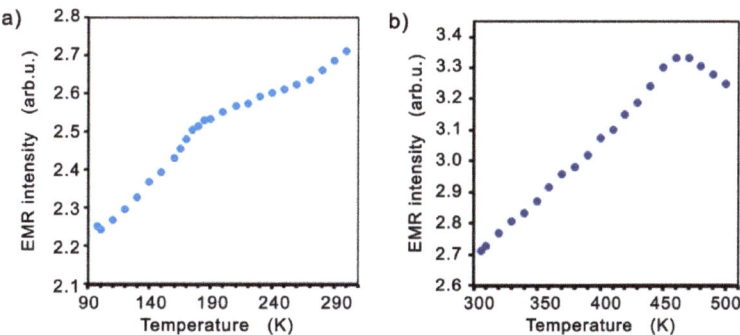

Figure 5. The EMR line intensities as a function of temperature in two temperature ranges, below (**a**), and close to T_C (**b**).

The nature of the temperature dependence of the line parameters (H_{pp}, I, and H_r) suggests that the T_C is higher than 500 K. The theory described by Becker [55] was used to analyze the temperature dependence (where T is the temperature of the measurement). Becker calculated the EMR resonance field shift and linewidth as a function of temperature and frequency near freezing temperature for spin-glass alloys, using RKKY exchange coupling and a smaller anisotropic interaction. To fit our line parameters' behavior, we adopted Becker's theory for the critical regime ($T \sim T_C$) and spin-glass regime ($T < T_C$). In $Zn_{0.8}Co_{0.2}O/Ag$, we can see an abnormal reduction in the linewidth resonance (H_{pp}) with a minimum at or near the critical temperature (Figure 6).

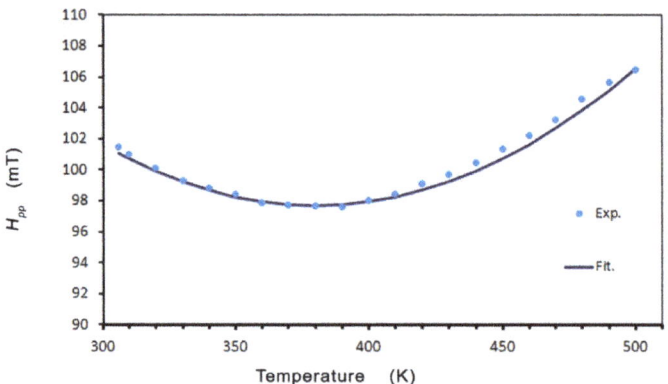

Figure 6. Temperature dependence of the resonance linewidth H_{pp}.

To fit the linewidth about the minimum, we used the function [50,51,55]:

$$\Delta H = a_0 + b' \left| \frac{T - T_{min}}{T_{min}} \right|^n \tag{1}$$

where a_0 is the residual linewidth; b' is the thermal broadening constant (independent of the orientation of the static field); n is the exponent of the expression for the length of the correlation associated with the distributed magnetization of Huber's theory of linewidth in isolated ferromagnets and antiferromagnets in the region near the critical temperature [56,57]; and T_{min} represents the temperature of the minimum linewidth. The large value of the exponent $n = 3\,v/2$ that we obtain is consistent with the presence of a strong perturbation, which is expected to increase the rate at which the correlation length decreases as the temperature moves away from the critical temperature. In the theory of the mean-field 3D Heisenberg model $v = 0.71$, and the value that we obtained was $v = 1.33$ [58]. The best fit of Equation (1) is shown in Figure 6 as the black line, where $a_0 = 97.7$ mT, $b' = 89$ mT, $T_{min} = 380$ K, and $n = 2.0$. This agrees with experimental data in the temperature range of 300 to 500 K, and the value of $a_0 = 97.7$ mT implies that the effects of the crystal field and demagnetization in $Zn_{0.8}Co_{0.2}O/Ag$ are high. The origin of the residual linewidth component in the spin-glass alloys has been assigned to local moment imperfections and crystal-field effects via a mechanism of demagnetization.

In the low-temperature regime ($T < T_C$), we can see a linewidth broadening and a shift in the resonance field. This is similar to the dependence in spin-glass alloys (for example, $AgMn_xSb_y$) and in molecule-based magnets [50,55]. In these alloys, the excess linewidth was assigned to an exchange-narrowed anisotropic interaction. With the decreasing temperature, the slowing down of the spin fluctuations reduces the effectiveness of the exchange-narrowing. Moreover, these alloys also show a related shift in the resonance field, which is neither a frequency-independent internal field nor a pure g-shift. Becker calculated the EMR linewidth and line shift effects for spin-glasses with anisotropy [55].

For systems with no remnant magnetization, Becker has shown that the resonance field and linewidth are given by formulas [50,51,55]:

$$\Delta H = \frac{ABT}{B^2 + T^2} \quad (2)$$

and

$$H_r = H_0 + \frac{AT^2}{B^2 + T^2} \quad (3)$$

where $A = \frac{g\mu_B K}{\hbar\omega\chi_\perp}$, $B = \frac{M_2}{Kk_b\omega}$, $H_0 = \frac{\hbar\omega}{g\mu_B}$; and ω is the resonance frequency. Here K is the constant of anisotropy, $\chi_\perp$ is the static susceptibility of transverse, and M_2 is associated with spin relaxation.

The best fits to the experimental data (shown in Figures 7 and 8) are A = 39.62 mT, B = 44.49 K, and H_0 = 19.12 mT. Very good agreement between the experimental data and the fitting (Equations (2) and (3)) is found. The linewidth and resonance field shift are compatible with a disordered ferrimagnet near and below critical temperature. Additional confirmation of the observed ferromagnetic properties is provided by the LFMA line. The absorption of microwave power centered at zero magnetic field has been reported in ferromagnetic materials and in various other materials, such as ferrites, high-temperature superconductors, and soft magnetic materials [44,46,59]. For soft magnetic materials, the LFMA signal is induced by low-field processes of spin magnetization [46]. The appearance of LMFA lines is an indicator of the ferromagnetic properties of the material. Figure 9 shows the magnetic hysteresis that is very often peeled off in the literature for LFMA lines.

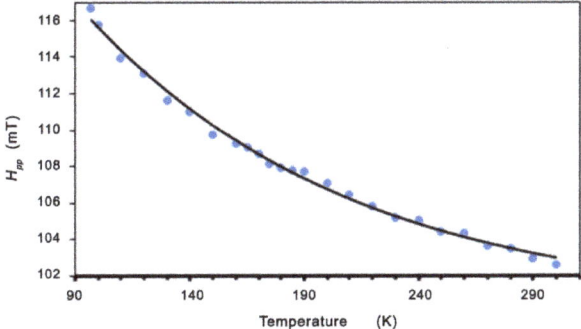

Figure 7. Temperature dependence of the peak-to-peak linewidth for $Zn_{0.8}Co_{0.2}O/Ag$. Theoretical line fitted based on Equation (2).

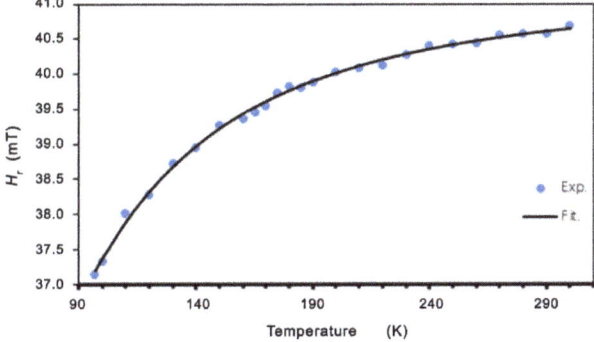

Figure 8. Temperature dependence of the resonance field for $Zn_{0.8}Co_{0.2}O/Ag$. Theoretical line fitted based on Equation (3).

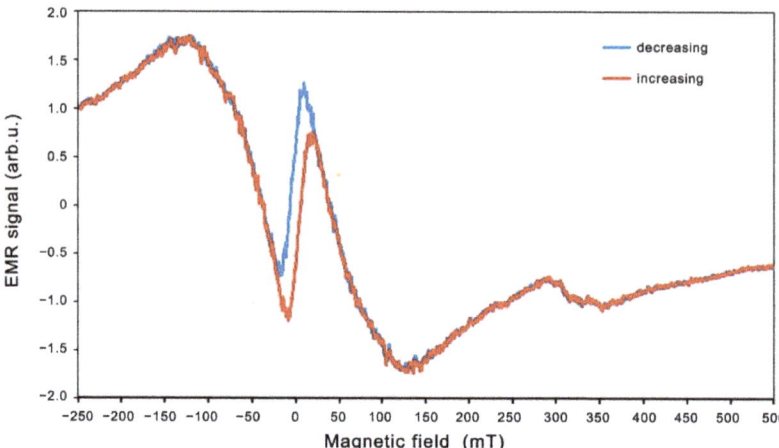

Figure 9. EMR spectrum recorded for an increasing (red) and decreasing (blue) magnetic field at 300 K for $Zn_{0.8}Co_{0.2}O/Ag$ on a quartz substrate.

A similar hysteresis of the LFMA line was observed for $Zn_{0.8}Co_{0.2}O/Ag$ on a silicon substrate, shown in Figure 10.

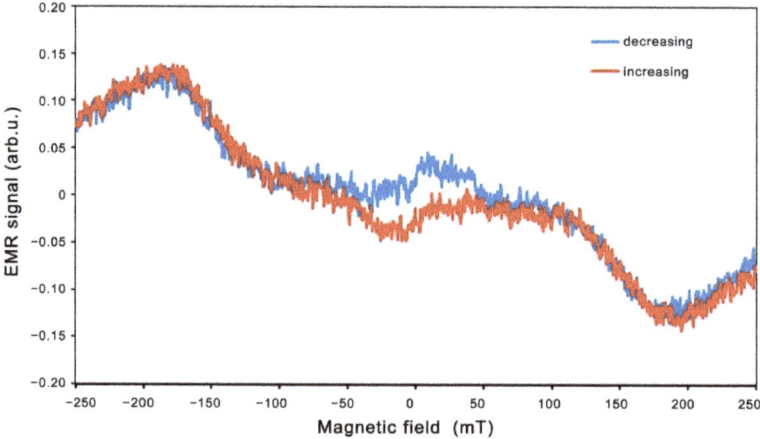

Figure 10. EMR spectrum recorded for an increasing (red) and a decreasing (blue) magnetic field at 140 K for $Zn_{0.8}Co_{0.2}O/Ag$ on a silicon substrate.

For the $Zn_{0.8}Co_{0.2}O/Ag$ film on a silicon substrate, we observe a lower intensity of the EMR lines and LFMA lines, but the nature of the observed magnetic properties is similar. The addition of gold produces a similar effect, although with a weaker intensity. The same layer deposited on a silicon substrate produces an analogous effect, including a hysteresis loop, although with a weaker intensity. The observed changes in the EMR spectrum obtained in both directions of registration show hysteresis only near the zero magnetic field; meanwhile in the rest of the range, the spectrum has an identical form and we do not observe changes in the shape of the EMR spectrum as presented in the paper [60].

3. Materials and Methods

Electron magnetic resonance measurements in the continuous wave X-band were taken on the Bruker FT-EPR ELEXSYS E580 spectrometer (Bruker Analytische Messtechnik,

Rheinstetten, Germany). To control the temperature, Bruker liquid nitrogen cryostats were used with the 41131 VT digital controller, and the angular dependences of the EPR spectra were performed using a one-degree programmable goniometer E218-1001 (Bruker Analytische Messtechnik, Rheinstetten, Germany).

Samples of ZnO doped with cobalt thin films ($Zn_{1-x}Co_xO$, $x = 0.2$) embedded with Ag NPs were obtained using a combination of PLD and homogeneous nucleation techniques. A homogeneous nucleation technique was chosen for the synthesis of silver nanoparticles (more details on the creation of noble metal NPs are given in the work [61]). A droplet of aqueous Ag NPs was then deposited on the surface of the substrate with further drying under surrounding conditions. For the PLD method, a silicon and quartz substrate was chosen. The $KGd(WO_4)_2$ laser was used—radiation characteristics: $\lambda = 1067$ nm, beam energy density 6–8 J/cm^2, repetition rate 10–0.3 Hz, pulse duration $t = 20$ ns. The technology module used the Q-switch to irradiate in modulated goodness factor mode. The deposition temperature (substrate temperature) was about 200 °C. The thickness of the resulting layer was about 300 nm. Therefore, a planar nanocomposite consisting of a $Zn_{0.8}Co_{0.2}O$ thin film deposited on silver nanoparticles ($Zn_{0.8}Co_{0.2}O/Ag$) was formed. The specificity of the PLD method is the wide particle energies spread, so we expect a deeper penetration of the applied particles into the substrate and the formation of a complex structure at the contact zone of the resulting layer and Ag NPs.

The basic parameters of the obtained layers were controlled, and the results are included in the Supplementary Materials.

4. Conclusions

We have performed X-band EMR studies of the $Zn_{0.8}Co_{0.2}O$ embedded with Ag NPs. A line in a broad asymmetric Dyson shape associated with magnetic interactions was observed. The results of a temperature dependence analysis for the EMR linewidth and resonant field based on Becker's model were developed. From the critical regime ($T \sim T_C$), the temperature $T_{min} = 380$ K was determined for the minimum width of the EMR linewidth. The value of $a_0 = 97.7$ mT suggests that demagnetization and the effects of the crystal field in $Zn_{0.8}Co_{0.2}O/Ag$ are high. For the range of the low-temperature regime ($T < T_C$), constants A and B related to the magnetic properties were determined. $A = 39.62$ mT, $B = 44.49$ K, and $H_0 = 19.12$ mT. Very good agreement between the experimental data and the fitting was found. The linewidth and resonance field shift are in accordance with a disordered ferrimagnet near and below the critical temperature. In addition, small, intense lines are assigned to the silver ions Ag^{2+} and Ag^0 [54]. There is a shift of the line towards the low-field direction for samples with silver nanoparticles.

The aim of this study was to investigate the effect of silver nanoparticles on the magnetic properties of $Zn_{0.8}Co_{0.2}O$. It was shown that the implementation of silver nanoparticles causes a significant shift of the ferromagnetic resonance (FMR) line (Figure 3) and that the ferromagnetic properties occur in a wide temperature range with an estimated Curie temperature above 500 K; the description is consistent with the adopted model, which can be considered as a major achievement of this work. Based on the results obtained, it can be claimed that a material with desirable magnetic properties at room temperature has been obtained, with potential applications that are important in spintronics.

Supplementary Materials: The following are available online at https://www.mdpi.com/article/10.3390/molecules27238500/s1, Figure S1: SEM images of the $Zn_{0.8}Co_{0.2}O/Ag$ film with clusters of Ag NPs; Figure S2: XRD diffractogram of $Zn_{0.8}Co_{0.2}O/Ag$ film on a silicon substrate; Figure S3: EDS analysis of the $Zn_{0.8}Co_{0.2}O/Ag$ composition; Table S1: The average crystallite size of Ag NPs; Table S2: EDS analysis of $Zn_{0.8}Co_{0.2}O/Ag$ composites

Author Contributions: Conceptualization, B.C. and I.S.; methodology, B.C. and I.S.; investigation, B.C., I.V. and R.V.G.; data curation, B.C., I.S. and I.R.; writing—original draft preparation, B.C. and I.S.; writing—review and editing, B.C., I.S. and I.V. All authors have read and agreed to the published version of the manuscript.

Funding: This research received no external funding.

Institutional Review Board Statement: Not applicable.

Informed Consent Statement: Not applicable.

Data Availability Statement: Data available in a publicly accessible repository.

Conflicts of Interest: The authors declare no conflict of interest.

Sample Availability: Samples of the $Zn_{0.8}Co_{0.2}O$/Ag compounds are available from the authors.

References

1. Pearton, S. *GaN and ZnO-Based Materials and Devices*; Pearton, S., Ed.; Springer Series in Materials Science; Springer: Berlin/Heidelberg, Germany, 2012; Volume 156, ISBN 978-3-642-23520-7.
2. Weng, J.; Zhang, Y.; Han, G.; Zhang, Y.; Xu, L.; Xu, J.; Huang, X.; Chen, K. Electrochemical deposition and characterization of wide band semiconductor ZnO thin film. *Thin Solid Films* **2005**, *478*, 25–29. [CrossRef]
3. Wilkinson, J.; Ucer, K.B.; Williams, R.T. Picosecond excitonic luminescence in ZnO and other wide-gap semiconductors. *Radiat. Meas.* **2004**, *38*, 501–505. [CrossRef]
4. Greuter, F. ZnO Varistors: From Grain Boundaries to Power Applications. In *Oxide Electronics*; John Wiley & Sons, Ltd.: Hoboken, NJ, USA, 2021; pp. 157–234.
5. Tsai, Y.T.; Chang, S.J.; Ji, L.W.; Hsiao, Y.J.; Tang, I.T.; Lu, H.Y.; Chu, Y.L. High Sensitivity of NO Gas Sensors Based on Novel Ag-Doped ZnO Nanoflowers Enhanced with a UV Light-Emitting Diode. *ACS Omega* **2018**, *3*, 13798–13807. [CrossRef]
6. Kang, Y.; Yu, F.; Zhang, L.; Wang, W.; Chen, L.; Li, Y. Review of ZnO-based nanomaterials in gas sensors. *Solid State Ion.* **2021**, *360*, 115544. [CrossRef]
7. Kolesnik, S.; Dabrowski, B.; Mais, J. Structural and magnetic properties of transition metal substituted ZnO. *J. Appl. Phys.* **2004**, *95*, 2582–2586. [CrossRef]
8. Samarth, N.; Furdyna, J.K. Electron paramagnetic resonance in $Cd_{1-x}Mn_xS$, $Cd_{1-x}Mn_xSe$, and $Cd_{1-x}Mn_xTe$. *Phys. Rev. B* **1988**, *37*, 9227–9239. [CrossRef]
9. Volbers, N.; Zhou, H.; Knies, C.; Pfisterer, D.; Sann, J.; Hofmann, D.M.; Meyer, B.K. Synthesis and characterization of ZnO:Co^{2+} nanoparticles. *Appl. Phys. A Mater. Sci. Process.* **2007**, *88*, 153–155. [CrossRef]
10. Awschalon, D.D.; Loss, D. *Semiconductors Spintronics and Quantum Computation*; Springer: Berlin/Heidelberg, Germany, 2002; ISBN 9783642075773.
11. Pearton, S.J.; Norton, D.P.; Heo, Y.W.; Tien, L.C.; Ivill, M.P.; Li, Y.; Kang, B.S.; Ren, F.; Kelly, J.; Hebard, A.F. ZnO spintronics and nanowire devices. *J. Electron. Mater.* **2006**, *35*, 862–868. [CrossRef]
12. Pan, F.; Song, C.; Liu, X.J.; Yang, Y.C.; Zeng, F. Ferromagnetism and possible application in spintronics of transition-metal-doped ZnO films. *Mater. Sci. Eng. R Rep.* **2008**, *62*, 1–35. [CrossRef]
13. Kisan, B.; Kumar, J.; Alagarsamy, P. Experimental and first-principles study of defect-induced electronic and magnetic properties of ZnO nanocrystals. *J. Phys. Chem. Solids* **2020**, *146*, 109580. [CrossRef]
14. Dietl, T.; Ohno, H.; Matsukura, F.; Cibert, J.; Ferrand, D. Zener Model Description of Ferromagnetism in Zinc-Blende Magnetic Semiconductors. *Science* **2000**, *287*, 1019–1022. [CrossRef] [PubMed]
15. Sato, K.; Katayama-Yoshida, H. Material design for transparent ferromagnets with ZnO-based magnetic semiconductors. *Jpn. J. Appl. Phys. Part 2 Lett.* **2000**, *39*, L555–L558. [CrossRef]
16. Sato, K.; Katayama-Yoshida, H. Stabilization of ferromagnetic states by electron doping in Fe-, Co- or Ni-doped ZnO. *Jpn. J. Appl. Phys. Part 2 Lett.* **2001**, *40*, L334–L336. [CrossRef]
17. Jung, S.W.; An, S.-J.; Yi, G.-C.; Jung, C.U.; Lee, S.-I.; Cho, S. Ferromagnetic properties of $Zn_{1-x}Mn_xO$ epitaxial thin films. *Appl. Phys. Lett.* **2002**, *80*, 4561–4563. [CrossRef]
18. Schwartz, D.A.; Norberg, N.S.; Nguyen, Q.P.; Parker, J.M.; Gamelin, D.R. Magnetic Quantum Dots: Synthesis, Spectroscopy, and Magnetism of Co^{2+}- and Ni^{2+}-Doped ZnO Nanocrystals. *J. Am. Chem. Soc.* **2003**, *125*, 13205–13218. [CrossRef] [PubMed]
19. Lee, H.-J.; Jeong, S.-Y.; Cho, C.R.; Park, C.H. Study of diluted magnetic semiconductor: Co-doped ZnO. *Appl. Phys. Lett.* **2002**, *81*, 4020–4022. [CrossRef]
20. Cho, Y.M.; Choo, W.K.; Kim, H.; Kim, D.; Ihm, Y. Effects of rapid thermal annealing on the ferromagnetic properties of sputtered $Zn_{1-x}(Co_{0.5}Fe_{0.5})_xO$ thin films. *Appl. Phys. Lett.* **2002**, *80*, 3358–3360. [CrossRef]
21. Sharma, P.; Gupta, A.; Rao, K.V.; Owens, F.J.; Sharma, R.; Ahuja, R.; Osorio-Guillen, J.M.; Johansson, B.; Gehring, G.A. Ferromagnetism above room temperature in bulk and transparent thin films of Mn-doped ZnO. *Nat. Mater.* **2003**, *2*, 673–677. [CrossRef]
22. Wakano, T.; Fujimura, N.; Morinaga, Y.; Abe, N.; Ashida, A.; Ito, T. Magnetic and magneto-transport properties of ZnO:Ni films. *Phys. E* **2001**, *10*, 260–264. [CrossRef]
23. Ueda, K.; Tabata, H.; Kawai, T. Magnetic and electric properties of transition-metal-doped ZnO films. *Appl. Phys. Lett.* **2001**, *79*, 988–990. [CrossRef]

24. Kim, J.H.; Kim, H.; Kim, D.; Ihm, Y.E.; Choo, W.K. Magnetic properties of epitaxially grown semiconducting $Zn_{1-x}Co_xO$ thin films by pulsed laser deposition. *J. Appl. Phys.* **2002**, *92*, 6066–6071. [CrossRef]
25. Ankiewicz, A.O.; Carmo, M.C.; Sobolev, N.A.; Gehlhoff, W.; Kaidashev, E.M.; Rahm, A.; Lorenz, M.; Grundmann, M. Electron paramagnetic resonance in transition metal-doped ZnO nanowires. *J. Appl. Phys.* **2007**, *101*, 024324. [CrossRef]
26. Jalbout, A.F.; Chen, H.; Whittenburg, S.L. Monte Carlo simulation on the indirect exchange interactions of Co-doped ZnO film. *Appl. Phys. Lett.* **2002**, *81*, 2217–2219. [CrossRef]
27. Lee, E.C.; Chang, K.J.J. Ferromagnetic versus antiferromagnetic interaction in Co-doped ZnO. *Phys. Rev. B-Condens. Matter Mater. Phys.* **2004**, *69*, 085205. [CrossRef]
28. Merga, G.; Cass, L.C.; Chipman, D.M.; Meisel, D. Probing silver nanoparticles during catalytic H_2 evolution. *J. Am. Chem. Soc.* **2008**, *130*, 7067–7076. [CrossRef] [PubMed]
29. Choi, J.S.; Jun, Y.W.; Yeon, S.I.; Kim, H.C.; Shin, J.S.; Cheon, J. Biocompatible heterostructured nanoparticles for multimodal biological detection. *J. Am. Chem. Soc.* **2006**, *128*, 15982–15983. [CrossRef]
30. Jeong, S.H.; Park, B.N.; Lee, S.B.; Boo, J.-H. Structural and optical properties of silver-doped zinc oxide sputtered films. *Surf. Coat. Technol.* **2005**, *193*, 340–344. [CrossRef]
31. Gouvêa, C.A.K.; Wypych, F.; Moraes, S.G.; Durán, N.; Peralta-Zamora, P. Semiconductor-assisted photodegradation of lignin, dye, and kraft effluent by Ag-doped ZnO. *Chemosphere* **2000**, *40*, 427–432. [CrossRef]
32. Georgekutty, R.; Seery, M.K.; Pillai, S.C. A highly efficient Ag-ZnO photocatalyst: Synthesis, properties, and mechanism. *J. Phys. Chem. C* **2008**, *112*, 13563–13570. [CrossRef]
33. Yin, X.; Que, W.; Fei, D.; Shen, F.; Guo, Q. Ag nanoparticle/ZnO nanorods nanocomposites derived by a seed-mediated method and their photocatalytic properties. *J. Alloys Compd.* **2012**, *524*, 13–21. [CrossRef]
34. Wang, J.; Fan, X.M.; Tian, K.; Zhou, Z.W.; Wang, Y. Largely improved photocatalytic properties of Ag/tetrapod-like ZnO nanocompounds prepared with different PEG contents. *Appl. Surf. Sci.* **2011**, *257*, 7763–7770. [CrossRef]
35. Zheng, Y.; Zheng, L.; Zhan, Y.; Lin, X.; Zheng, Q.; Wei, K. Ag/ZnO heterostructure nanocrystals: Synthesis, characterization, and photocatalysis. *Inorg. Chem.* **2007**, *46*, 6980–6986. [CrossRef] [PubMed]
36. Lin, D.; Wu, H.; Zhang, R.; Pan, W. Enhanced photocatalysis of electrospun Ag-ZnO heterostructured nanofibers. *Chem. Mater.* **2009**, *21*, 3479–3484. [CrossRef]
37. Zhang, D.; Liu, X.; Wang, X. Growth and photocatalytic activity of ZnO nanosheets stabilized by Ag nanoparticles. *J. Alloys Compd.* **2011**, *509*, 4972–4977. [CrossRef]
38. Hu, H.; Wang, Z.; Wang, S.; Zhang, F.; Zhao, S.; Zhu, S. ZnO/Ag heterogeneous structure nanoarrays: Photocatalytic synthesis and used as substrate for surface-enhanced Raman scattering detection. *J. Alloys Compd.* **2011**, *509*, 2016–2020. [CrossRef]
39. Aguirre, M.E.; Rodríguez, H.B.; San Román, E.; Feldhoff, A.; Grela, M.A. Ag@ZnO Core–Shell Nanoparticles Formed by the Timely Reduction of $Ag+$ Ions and Zinc Acetate Hydrolysis in N,N-Dimethylformamide: Mechanism of Growth and Photocatalytic Properties. *J. Phys. Chem. C* **2011**, *115*, 24967–24974. [CrossRef]
40. Wang, C.C.; Shieu, F.S.; Shih, H.C. Ag-nanoparticle enhanced photodegradation of ZnO nanostructures: Investigation using photoluminescence and ESR studies. *J. Environ. Chem. Eng.* **2021**, *9*, 104707. [CrossRef]
41. Yan, Y.; Al-Jassim, M.M.; Wei, S.H. Doping of ZnO by group-IB elements. *Appl. Phys. Lett.* **2006**, *89*, 181912. [CrossRef]
42. Ahn, B.D.; Kang, H.S.; Kim, J.H.; Kim, G.H.; Chang, H.W.; Lee, S.Y. Synthesis and analysis of Ag-doped ZnO. *J. Appl. Phys.* **2006**, *100*, 093701. [CrossRef]
43. Brauer, G.; Kuriplach, J.; Ling, C.C.; Djurišić, A.B. Activities towards p-type doping of ZnO. *J. Phys. Conf. Ser.* **2011**, *265*, 012002. [CrossRef]
44. Srinivasu, V.V.; Lofland, S.E.; Bhagat, S.M.; Ghosh, K.; Tyagi, S.D. Temperature and field dependence of microwave losses in manganite powders. *J. Appl. Phys.* **1999**, *86*, 1067–1072. [CrossRef]
45. Alvarez, G.; Zamorano, R. Characteristics of the magnetosensitive non-resonant power absorption of microwave by magnetic materials. *J. Alloys Compd.* **2004**, *369*, 231–234. [CrossRef]
46. Montiel, H.; Alvarez, G.; Betancourt, I.; Zamorano, R.; Valenzuela, R. Correlations between low-field microwave absorption and magnetoimpedance in Co-based amorphous ribbons. *Appl. Phys. Lett.* **2005**, *86*, 072503. [CrossRef]
47. Stefaniuk, I.; Cieniek, B.; Virt, I.S. Magnetic Properties of Zinc-Oxide Composite Doped with Transition Metal Ions (Mn, Co, Cr). *Curr. Top. Biophys.* **2010**, *33*, 221–226.
48. Cieniek, B.; Stefaniuk, I.; Virt, I.S. EPR study of ZnO:Co thin films grown by the PLD method. *Nukleonika* **2013**, *58*, 359–363.
49. Cieniek, B.; Stefaniuk, I.; Virt, I.S. EMR spectra thin films doped with high concentration of Co and Cr on quartz and sapphire substrates. *Acta Phys. Pol. A* **2017**, *132*, 30–33. [CrossRef]
50. Mozurkewich, G.; Elliott, J.H.; Hardiman, M.; Orbach, R. Exchange-narrowed anisotropy contribution to the EPR width and shift in the Ag-Mn spin-glass. *Phys. Rev. B* **1984**, *29*, 278–287. [CrossRef]
51. Long, S.M.; Zhou, P.; Miller, J.S.; Epstein, A.J. Electron Spin Resonance Study of The Disorder in The V(TCNE)x.y(MeCN) High-Tc Molecule-Based Magnet. *Mol. Cryst. Liq. Cryst. Sci. Technol. Sect. A Mol. Cryst. Liq. Cryst.* **1995**, *272*, 207–215. [CrossRef]
52. Yıldız, F.; Yalçm, O.; Özdemir, M.; Aktaş, B.; Köseoğlu, Y.; Jiang, J.S. Magnetic properties of Sm-Co/Fe exchange spring magnets. *J. Magn. Magn. Mater.* **2004**, *272*, E1941–E1942. [CrossRef]
53. Yaln, O. Ferromagnetic Resonance. In *Ferromagnetic Resonance-Theory and Applications*; IntechOpen: London, UK, 2013; ISBN 978-953-51-1186-3.

54. Mandal, A.R.; Mandal, S.K. Electron spin resonance in silver-doped PbS nanorods. *J. Exp. Nanosci.* **2010**, *5*, 189–198. [CrossRef]
55. Becker, K.W. Theory of electron-spin-resonance linewidth and line-shift effects in spin-glasses with anisotropy and zero remanent magnetization. *Phys. Rev. B* **1982**, *26*, 2409–2413. [CrossRef]
56. Huber, D.L. EPR linewidths in RbMnF3 and MnF2. *Phys. Lett. A* **1971**, *37*, 283–284. [CrossRef]
57. Zomack, M.; Baberschke, K.; Barnes, S.E. Magnetic resonance in the spin-glass (LaGd)Al2. *Phys. Rev. B* **1983**, *27*, 4135–4148. [CrossRef]
58. Yeomans, J.M. *Statistical Mechanics of Phase Transitions*; Clarendon Press: Oxford, UK, 1992; Volume 19, ISBN 9780198517306.
59. Bhat, S.V.; Ramakrishnan, T.V.; Ganguly, P.; Rao, C.N.R. Absorption of electromagnetic radiation by superconducting $YBa_2Cu_3O_7$: An oxygen-induced phenomenon. *J. Phys. C Solid State Phys.* **1987**, *20*, L559–L563. [CrossRef]
60. Gafurov, M.R.; Kurkin, I.N.; Kurzin, S.P. Inhomogeneity of the intrinsic magnetic field in superconducting $YBa_2Cu_3O_X$ compounds as revealed by a rare-earth EPR probe. *Supercond. Sci. Technol.* **2005**, *18*, 1183–1189. [CrossRef]
61. Savchuk, V.V.; Gamernyk, R.V.; Virt, I.S.; Malynych, S.Z.; Pinchuk, A.O. Plasmon-exciton coupling in nanostructured metal-semiconductor composite films. *AIP Adv.* **2019**, *9*, 045021. [CrossRef]

Article

Electron Paramagnetic Resonance Study of PbSe, PbTe, and PbTe:In Semiconductors Obtained by the Pulsed Laser Deposition Method

Aleksandra Wędrychowicz [1,*], Bogumił Cieniek [2], Ireneusz Stefaniuk [2], Ihor Virt [3] and Romana Śliwa [4]

1. Doctoral School of Engineering and Technical Sciences, Rzeszow University of Technology, Powstańców Warszawy 12, 35-959 Rzeszów, Poland
2. Institute of Materials Engineering, College of Natural Sciences, University of Rzeszow, Pigonia 1, 35-310 Rzeszow, Poland; bcieniek@ur.edu.pl (B.C.); istef@ur.edu.pl (I.S.)
3. Institute of Physics, College of Natural Sciences, University of Rzeszow, Pigonia 1, 35-310 Rzeszow, Poland; ivirt@ur.edu.pl
4. Faculty of Mechanical Engineering and Aeronautics, Rzeszow University of Technology, Powstańców Warszawy 12, 35-959 Rzeszów, Poland; rsliwa@prz.edu.pl
* Correspondence: wedrychowicz.aleksandra@gmail.com

Abstract: The magnetic properties of lead selenide (PbSe) and indium-doped lead telluride (PbTe:In) composites have been studied by using the electron paramagnetic resonance (EPR) technique. The samples were obtained by using the pulsed laser deposition method (PLD). Temperature dependences of the EPR spectra were obtained. The analysis of the temperature dependencies of the integral intensity of the EPR spectra was performed using the Curie–Weiss law. In these materials, the paramagnetic centers of Pb^{1+} and Pb^{3+} ions were identified. The results are discussed.

Keywords: EPR; PbTe; PbSe; DMS; paramagnetic species

1. Introduction

The great interest in ferromagnetic semiconductors with a wide band gap [1,2] is related to the possibilities of their various applications, e.g., in spintronics and so-called translucent electronics [3–5]. The longitudinal optical phonons are the principal scattering mechanism in p-type PbTe. On the contrary, the scattering caused by transverse optical mechanisms is the weakest [6]. The density approximation results in an energy gap, the experimental value of which is greater than in most materials. The fitting of the experimental range was proposed using the Slater–Koster fitting method. Only the addition of a compression component to the narrowed Hamiltonian definition must result in significant modifications in the band structure [7]. The main difficulty in the production of materials with low thermal conductivity is the problem with the effective scattering of phonons in the entire frequency spectrum. Some calculations show that the passage of PbTe to a ferroelectric phase transition can provide a balanced solution to this problem [8]. In recent years, the topic of functional ferromagnetic semiconductors has aroused great interest. The ferromagnetic reaction usually takes place above room temperature in thin layers of semiconductors and oxides doped with small amounts of magnetic compounds [9–11]. EPR studies have proven that Cr^{3+} ions describe the n-type conductivity, as well as the magnetic properties of the PbCrTe compound. The paramagnetic resonance and its correlation with the chromium concentration prove the existence of a Cr donor resonance in PbTe. There is a noticeable shift in the g factor, which is caused by the increased concentration of the carrier. It is the earliest manifestation of the existence of a large sp-d coupling between Cr^{3+} ions and conducting electrons [12]. All samples show the dependence of the Curie temperature (T_c) on the carrier concentration p, and it has the form of a threshold [13]. The result of the displacement of factor g is that the magnetic moment Mn^{2+} of the semiconductor diagrams PbTe and SnTe is observed for n- and p-type crystals. The observation of this effect makes it possible to determine the carrier for holes and electrons in PbTe and light and heavy holes in SnTe [14]. Although PbTe is a reference thermoelectric material, its applications are still being investigated by attempts to change its properties [15]. There are models of thermoelectric transport mechanisms in the L and ΣPbTe valleys, with particular emphasis on thermally induced shifts. Semiconductors with magnetic properties are an interesting research object due to their spintronic properties and the interaction of spin–spin exchanges between localized magnetic moments and band electrons [16]. PbTe with an energy gap of 0.29 eV at a temperature of 300 K [17] are used for infrared detectors and solar cells [18–20]. Based on $Bi_2(Te, Se)_3$, Ag_2Te, PbTe, and SnSe, functional fibers can be obtained as new flexible materials for thermoelectric devices [21,22]. Furthermore, there are potential applications of thermoelectric fibers or devices for electricity generation [23]. Due to the metallic conductivity of composites, the skin effect limits the penetration depth of microwave radiation to approximately 10 μm, resulting in an asymmetrical (dissonant) shape of the resonance lines [24]. The most interesting effects related to the influence of electronic properties of semi-magnetic semiconductors on their magnetic behavior are observed in the IV–VI groups of semiconductors, such as PbSe or PbTe [25–29]. Interesting ferromagnetic properties have been observed in many diluted magnetic semiconductors (DMS), e.g., ZnTe + Mn [30] and CdTe + Cr [31]. In addition, interesting ferromagnetic properties have been studied in works [32–35] for the doping of the Cu ion in ZnO. Although Cu, CuO, and Cu_2O are not ferromagnetic, typical ferromagnetic interactions of DMS are observed. Depending on the oxidation state of the copper ion, paramagnetic Cu^{2+}, and diamagnetic Cu^+, we can "switch" the magnetic properties by interactions with the defects. The introduction of hydrogen (H^+) into the (Zn, Cu)O layer results in the appearance of ferromagnetic properties because of the interaction with the defects. Curie temperatures of 42.5 K (spin-only) and 106.1K (spin–orbit coupling), respectively, were observed, depending on the type of defect and associated interactions [35]. A high Curie temperature ($T_C \approx 320$ K) and band gap opening are two key challenges ($MnSe_2$) for the integration of photoconductivity into two-dimensional (2D) magnets [36]. The primary subject of this paper is a review of the magnetic properties of the PbSe, PbTe, and PbTe:In layers studied by the EPR method, learning about these magnetic properties in combination with electrical measurements can be helpful in building thermoelectric devices.

2. Results and Discussion

EPR Measurements

X-band EPR spectra were measured for all samples, Figure 1 shows the spectra at room temperature (RT).

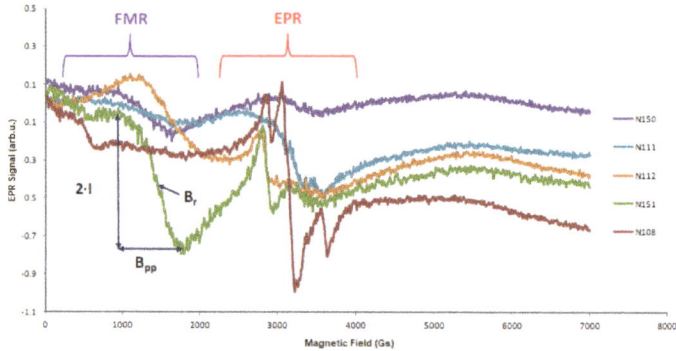

Figure 1. EPR measurements of PbTe (N111), PbTe (N112), PbTe (N150), PbTe:In (N151), and PbSe (N108) at room temperature. FMR-line area for ferromagnetic resonance, EPR- line area for paramagnetic resonance, B_{pp}-peak-to-peak width of the resonance line, I-resonance line intensity, and B_r-resonance field.

All spectra are characterized by an EPR line in the vicinity of 300 mT. For samples N151, N112, and N150, a wide line was observed in a low field of about 150 mT. However, in the case of the EPR spectrum for PbSe, we observed a typical Pb^{2+} ion line with a visible hyperfine structure in the vicinity of 300 mT. The configuration of the outer electron shell of Pb is $6s^2 6p^2$ and, thus, S = 1, L = 1, and J = L-S $\neq$ 0. Therefore, the conditions enabling the prevalence of polarization para-magnetism of an atomic electron shell over its precession diamagnetism are fulfilled. On the other hand, in the PbTe interstitial, the Pb atom will be surrounded by alternatively positioned ions of opposite signs, Pb^{2+} and Te^{2-}, forming a PbTe lattice of the NaCl type [1]. As a result, the outer electron shell of interstitial Pb will experience a strong attraction to Pb^{2+} ions of the metal sublattice and a strong repulsion from Te^{2-} of the chalcogen one. Thus, the interstitial Pb in the PbTe lattice can be a strong Van Vleck-type paramagnetic center. The Pb^{2+} ion has the electron configuration $6s^2$ and therefore has no electron magnetic moment and no observed EPR spectra. Lead ions, Pb^+ and Pb^{3+}, have electronic configurations $6s^2 p^1$ and $6s^1$, respectively. In the EPR spectra of both ions, a hyperfine structure induced by the interaction of an unpaired electron with the magnetic moment of the 207Pb nucleus (spin l = 1/2, natural content 21.1%) should be observed. The natural occurrence of even Pb isotopes without a nuclear spin (I = 0) is 78.9%. EPR spectra for Pb ions have been observed in many works, (see, e.g., [37–41]). They are successfully described using a rhombic-symmetry spin Hamiltonian (1) with an electron spin S = 1/2 and nuclear spins I = 0 or $\frac{1}{2}$ of the form (in the usual notation).

$$H_s = g\beta BS + AIS \quad (1)$$

where: g—spectroscopic splitting factor, β-Bohr magneton constant, B-magnetic field, A-hyperfine structure tensor, I-nucleus spin, and S-total spin.

Based on this relationship, the factor g_{eff} values were determined (see Table 1), while, for PbSe, the hyperfine structure constant A. Moreover, we performed fitting and simulation using MATLAB software with an Easyspin toolbox for both Pb^{3+} and Pb^{1+} ions at room temperature. The obtained results are shown in Figure 2. The parameters of the spine Hamiltonian (Equation (1)) were determined. For the Pb^{1+} ion, g = (1.000, 1.002, 1.504) and A = (5.20, 3.62, 4.58) MHz (N108); for the Pb^{3+} ion, g = (2.105, 2.106, 2.105), A = (1584.0, 1579.7, 1757.0) MHz (N150), and a g = (2055, 2006, 2216) and A = (1634.0, 1673.6, 1754.0) MHz (N150). In the literature (e.g., [41]), for Pb^+, g_{eff} = 1.12, and, for Pb^{3+}, $g_{eff} \approx 2$. We can assume that we mainly observed the EPR spectrum from Pb^{3+} ions and, for 3 samples, additional magnetic interactions.

Table 1. Summary of the Curie temperature with constant and g_{eff} parameters from EPR measurements at room temperature.

Sample	Magnetic Properties		g_{eff} (RT)		
	T_c(K)	C	"Ferromagnetic" Line ($g_{eff} \approx 5$)	Pb^{3+} ($g_{eff} \approx 2$)	Pb^{1+} ($g_{eff} \approx 1.1$)
N108	-	-		2.14	
N150	75.83	3.31×10^7	5.20	2.07	1.115
N151	132.99	1.94×10^6	4.67	2.36	
N112	5.31	1.79×10^8	4.46	2.39	
N111	174.21	1.59×10^{10}	-	2.11	

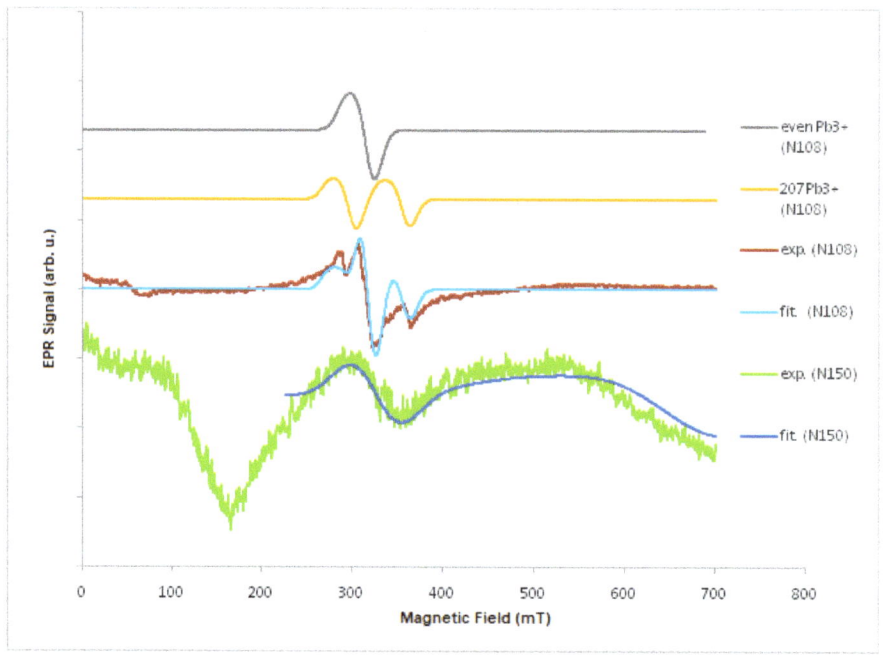

Figure 2. EPR spectra at room temperature for PbTe (N150) and PbSe (N108) experimental (exp.) and fitted (fit), as well as an example of simulated EPR spectra for 207Pb^{3+} ion and even Pb isotopes (206Pb, 208Pb, and 210Pb).

To characterize the samples from a magnetic point of view, we made dependencies of the EPR spectra as a function of the temperature. Figures 1 and 3–5 present the EPR spectra.

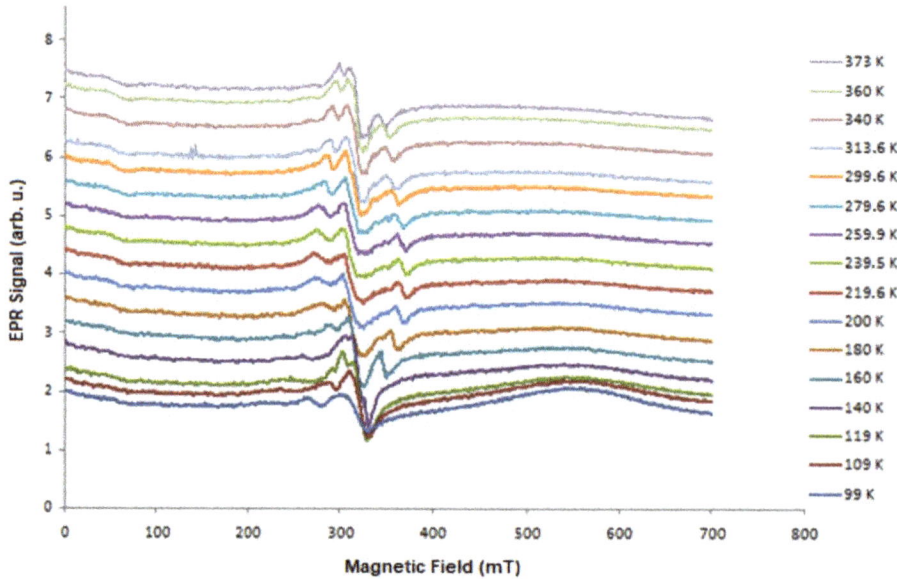

Figure 3. Temperature dependence of the PbSe EPR signal (N108).

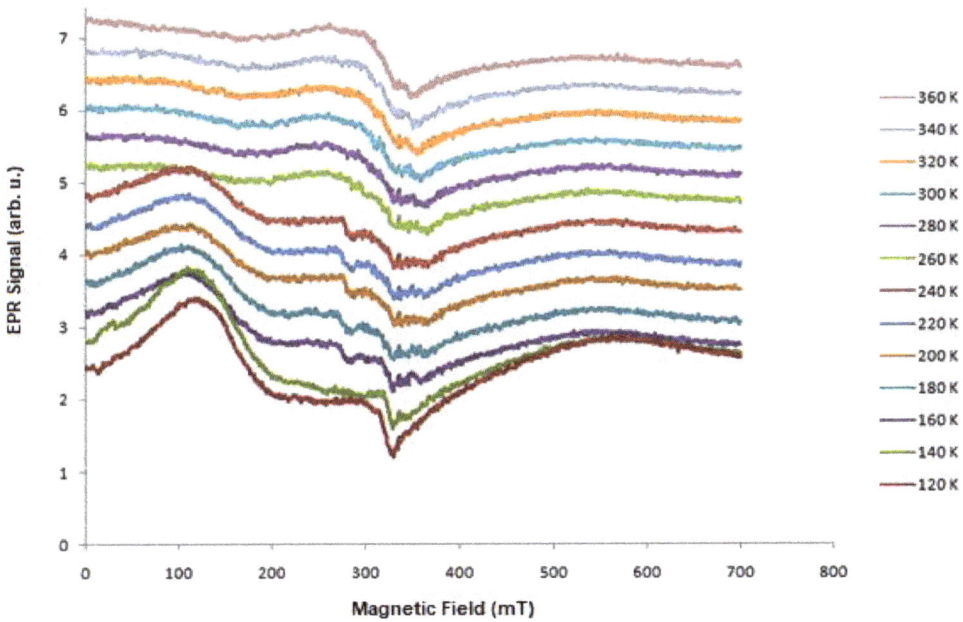

Figure 4. Temperature dependence of the PbTe EPR signal (N111).

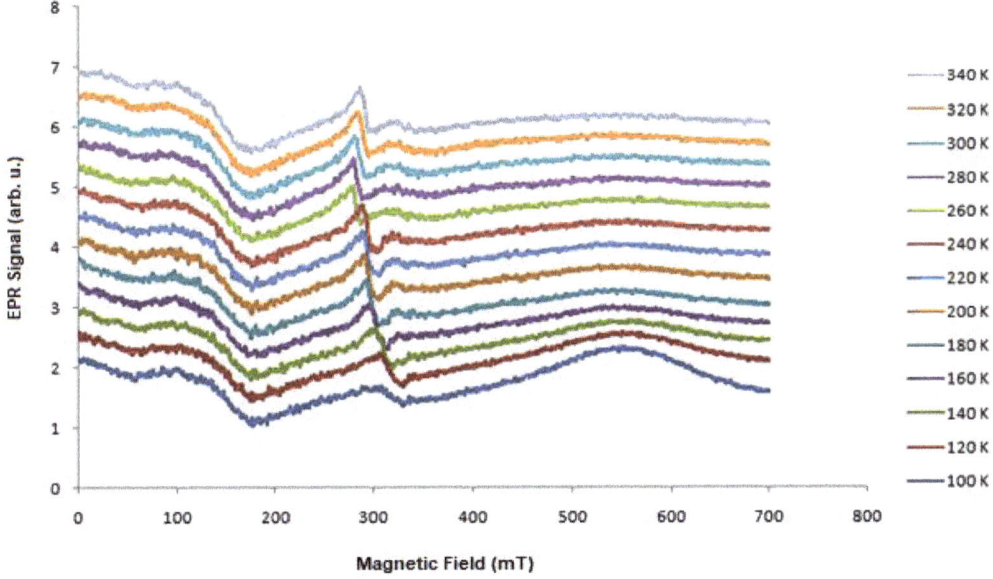

Figure 5. Temperature dependence of the EPR signal of PbTe:In (N151).

On the basis of the performed EPR measurements, the line parameters were determined, peak-to-peak width (B_{pp}), line intensity (I), and resonant field (B_r). From the dependence (1), the value of the factor g_{eff} was determined. Figures 6–8 show the dependence of g_{eff} as a function of temperature for the "ferromagnetic" line ($g_{eff} \approx 5$), the Pb^{3+} ($g_{eff} \approx 2$), and Pb^{1+} ($g_{eff} \approx 1.1$). For the EPR line with $g_{eff} \approx 1.1$, we observe the overlapping of signals from the Pb^{1+} ion and various defects related to oxygen ions [42]. At lower temperatures, there is a clear separation of these lines, and for the Pb^{1+} ion, we can observe a clear and narrow line (e.g., sample N150), while, in the other samples, this line is much weaker.

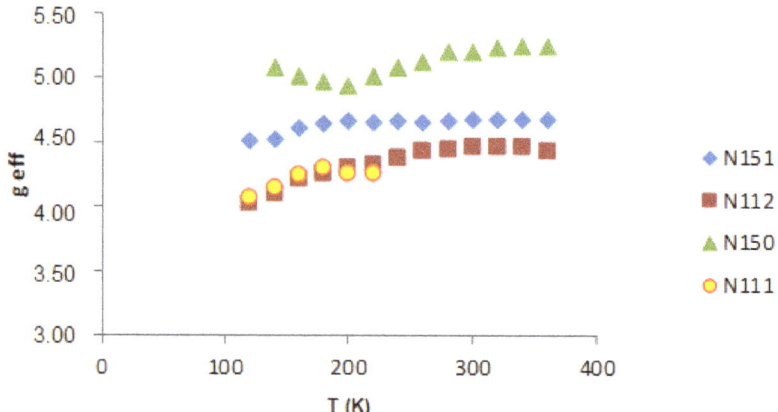

Figure 6. g_{eff} as a function of the temperature for the "ferromagnetic" line ($g_{eff} \approx 5$).

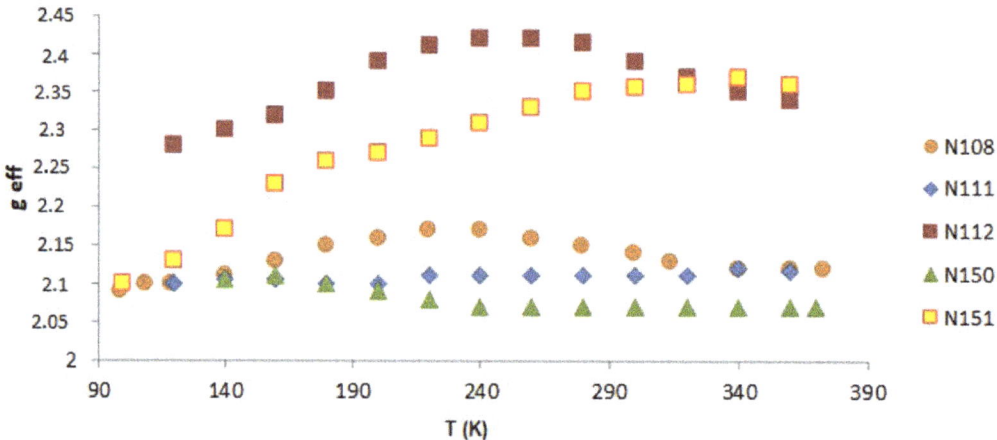

Figure 7. g_{eff} as a function of the temperature for Pb^{3+} ($g_{eff} \approx 2$).

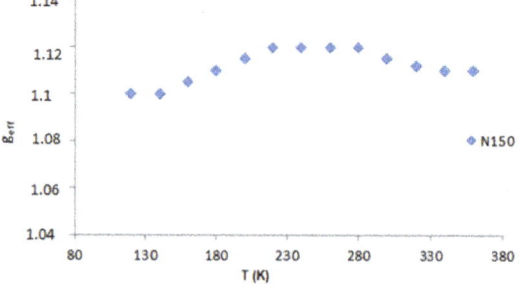

Figure 8. g_{eff} as a function of the temperature for Pb^{1+} ($g_{eff} \approx 1.1$).

The greatest differences are observed for the "ferromagnetic" line, whereas the g_{eff} value for the Pb^{3+} ion is constant in this temperature range.

An analysis of the changes in the EPR spectrum as a function of temperature on the total intensity of the EPR spectrum was performed. The total intensity of the EPR spectrum is determined on the basic line parameters according to the relation:

$$I_{EPR} = I(B_{pp})^2 \qquad (2)$$

We use the Curie–Weiss law to analyze the temperature dependence of the integral intensity, which is directly proportional to the magnetic susceptibility χ. A linear increase of $\chi^{-1}(T)$ at higher temperatures can be fitted to the Curie–Weiss law:

$$(\chi - \chi_0)^{-1}(T) = \frac{T}{C} - \frac{T_C}{C} \qquad (3)$$

where C is the Curie constant, Tc is the paramagnetic Curie temperature, and χ_0 is a temperature-independent term to account for the diamagnetic host and any Pauli Paramagnetism contribution. An example of the relationship $(\chi - \chi_0)^{-1}(T)$ for the N112 sample is shown in Figure 9. The calculated values of the Curie temperature and the Curie constant for all samples are presented in Table 1. The lines are linear extrapolations illustrating the ferromagnetic (positive) Curie-Weiss temperatures.

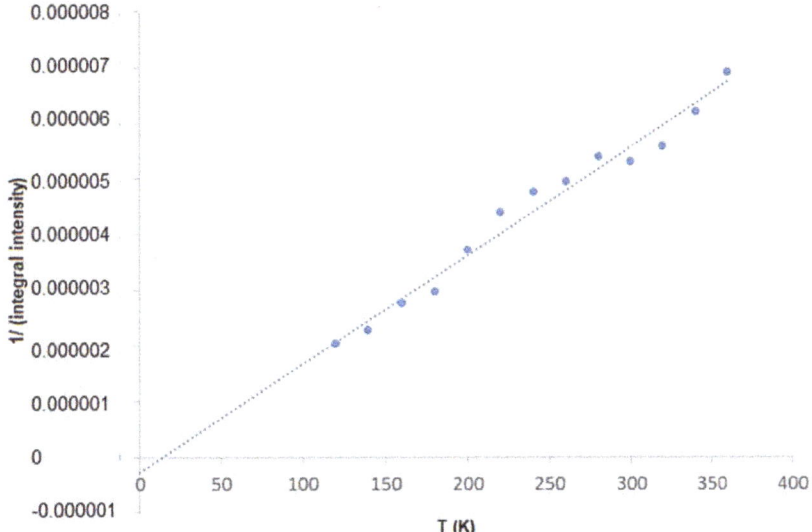

Figure 9. Temperature dependence of the 1/(integral intensity) of PbTe (N112).

When analyzing the obtained results, depending on the parameters of the obtained layers, we notice that ferromagnetic interactions appeared in only four samples. For sample N112, we observed a weak "ferromagnetic" line, and it was in correlation with the low Curie temperature. For sample N111, we observed the appearance of lines from ferromagnetic interactions at a temperature of about 240 K.

The highest ferromagnetic interactions were observed for PbTe (N111) and indium-doped PbTe (N151). In addition, the substrate temperature for the range of 200–250 K has a large influence, the PbTe layers have ferromagnetic properties, and, for a lower temperature, we do not observe ferromagnetic properties, while, at 300 K, these properties are still visible but weaker, and the obtained Curie temperature is lower. The appearance of the low-field line (LFMA) together with the "ferromagnetic" line is interesting; especially, it is visible in Figure 4 (N111). Different interpretations and explanations have been presented to try and explain the appearance of the LFMA signal at B = 0 in a wide variety of materials. For magnets and ferrites, the LFMA signal is associated with the onset of the ordered phase and provides a sensitive detector of magnetic ordering [43,44]. For soft magnetic materials, the LFMA signal is due to low-field spin magnetization processes [45]. In our case, we connected the appearance of the LFMA signal with the occurrence of ferromagnetic properties.

3. Materials and Methods

The experimental setup of the pulsed laser deposition method (PLD) used to deposit the PbSe and PbTe layers has previously been extensively described [46] and is only briefly described here. It uses a Q-switched Nd^{3+}: $KGd(WO_4)_2$ laser (λ = 1067 nm, pulse duration τ = 20 ns, 6–8 J/cm^2 fluence, and 0.3 Hz repetition rate) to ablate the polycrystalline targets in a quartz steel chamber. The final pressure in the deposition chamber was in the low 10^{-6} Pa. The layers were deposited under residual vacuum. For various samples, the temperature of the substrate (T_s) was changed, as well as the layer deposition time, related to the number of pulses and, hence, their thickness. The layers were prepared on quartz for EPR measurements (Table 2) and Al_2O_3 for electrical measurements (Table 3).

Table 2. Layer growth parameters deposited on a quartz substrate for EPR measurements.

	N111	N112	N150	N151	N108
	PbTe	PbTe	PbTe	PbTe:In	PbSe
T_s(°C)	120	200	300	250	220
Time (min)	20	25	25	30	25
Number of pulses	400	500	500	600	500

Table 3. Layer growth parameters deposited on the Al_2O_3 substrate for electrical measurements.

	N110	N158	N160
	PbTe	PbTe:In	PbSe
T_s(°C)	200	200	200
Time (min)	25	45	45
Number of pulses	300	550	550

EPR measurements were performed in X-band (~9.5 GHz) using a Bruker EleXSYS-E580 spectrometer (Billerica, MA, USA and Karlsruhe, Germany) equipped with a Bruker liquid Ngas flow cryostat with the 41131 VT digital controller (Bruker Analytische Messtechnik, Rheinstetten, Germany) within the temperature range 100–400 K.

Measurements of the electrical parameters of the semiconductor layers were performed on the developed automated installation developed according to the classical method. After applying a voltage to the sample (10 V), the current flowing is measured. During the measurement, the sample (layer) was placed in a standard copper-based holder with four measuring probes and a built-in reference resistor for current measurement with a digital microvoltmeter. The handle is attached through a detachable joint in the middle of a glass cylinder, in which a precise temperature sensor is mounted on a thermocouple connection. The production of reliable ohmic contacts that do not damage the layer and meet all the requirements was carried out by pre-brazing indium traces on the surface of the Al_2O_3 substrate prior to the production of the layer.

The structural quality of the respective films was investigated by transmission high-energy electron diffraction (THEED) for the samples grown on KCl (001) substrates. We used an EMR-100 electron diffractometer with acceleration voltages of 60–80 kV. For a better explanation of the good crystalline quality of the samples, in Figure 10, a summary of THEED images (a) for the PbTe:In layer studied and (b) PbTe published in Figure 10a) in [33]. All samples in Table 2 are of similar quality, as can be seen in the THEED image in Figure 10a. However, the same layers with different thicknesses are described in [33], and the parameters of the elemental cell are determined for them, along with the assigned Muller indexes. For example, the PbTe layer (Figure 10b) from Figure 11 [33].

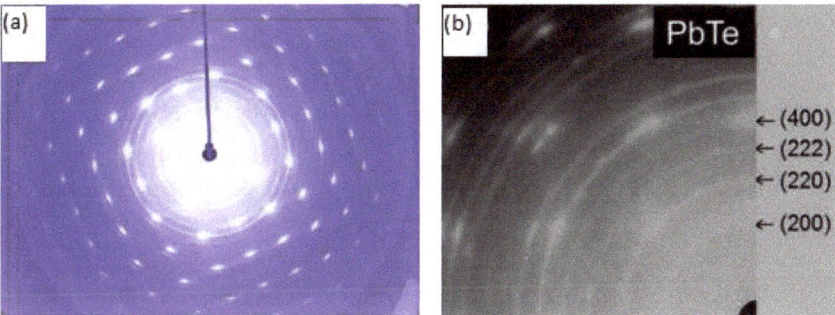

Figure 10. THEED patterns of (**a**) PbTe:In and (**b**) PbTe. The most pronounced Miller indices (hkl) are indicated.

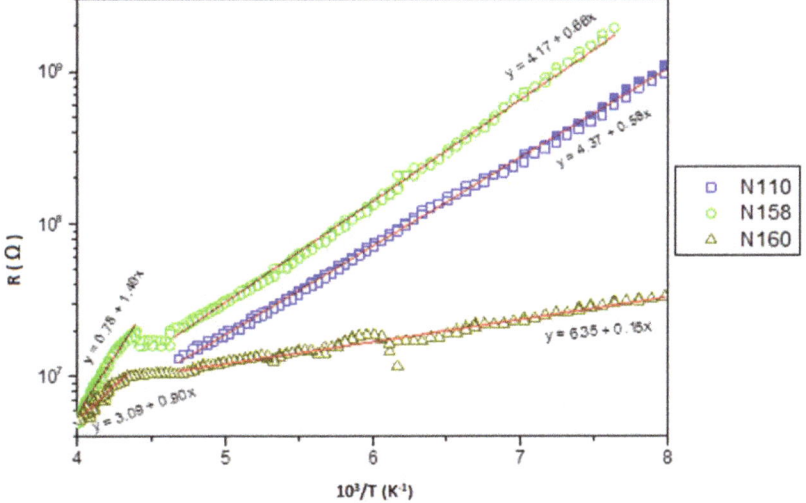

Figure 11. Electrical resistance as a function of the temperature of PbTe (N110), PbTe:In (N158), and PbSe (N160).

Electrical Measurements

The properties of the established contacts were controlled by analyzing the volt-ampere characteristics of the samples. The measurement temperature range was 77–500 K (sample temperature change).

Measurement soft electrical resistance as a function of temperature are shown in Figure 11. The type (carrier type) of conductivity is defined by the sign of the thermoelectromotive force (t_{ers}). The electrical characteristics of the layers are presented in the following coordinates: resistivity to inverse temperature ($\rho\ 10^3/T$).

The dependence of the resistance as a function of the temperature is linear over a wide range for all samples, and we see only differences in the slope of this line. On the other hand, for both samples, PbSe and PbTe:In at about 220 K, we observe a change in the slope of the straight line. In all samples, two drops in activation energy, depending on the resistivity temperature, are related to self-conductivity (at relatively high temperatures) and to doped conductivity (at temperatures below 200 K).

In paper [33], THEED images with high reorientation of crystallites were observed, practically only Debye-Scherrer rings and weakly visible reflections were visible. On the other hand, the RHEED images presented in Figure 10 are characterized by very good reflections with weakly visible Debye-Scherrer rings.

4. Conclusions

Electrical measurements were carried out to indicate that the obtained materials are of a semiconductor nature. All samples observed two activation energy slopes on the temperature dependence of resistivity: related to self-conductivity at relatively high temperatures and to a doped conductivity at temperatures below 200 K.

Identification of the paramagnetic centers present in the tested materials was carried out using EPR measurements. It was determined that the EPR spectrum comes from the Pb^{1+} ions in the sample (it is most visible in sample N150) and from Pb^{3+}. The g_{eff} values were found, which agree well with the data reported in the literature. In addition, we observed additional defects, i.e., vacancies and interstitial ions, which cause a widening of the line for $g_{eff} \approx 2$ and $g_{eff} \approx 1.1$. The natural oxidation state of the Pb ion is Pb^{2+}, while our measurements indicate the presence of both Pb^{3+} ($g_{eff} \approx 2$) and Pb^{1+} ($g_{eff} \approx 1.1$) ions. The EPR line derived from Pb^{1+} ions was weaker, due to the large broadening of the line and, thus, lower intensity. The observation of these two ions is related to the charge compensation in all samples. The Pb^{2+} ions are probably responsible for the ferromagnetic interactions ($g_{eff} \approx 5$). Pb^{1+} ions are observed at low temperatures in all samples.

A hyperfine structure for PbSe was observed. The determined temperature dependence was almost linear, while the constant value A = 1740 MHz (at 109 K) changed as a function of the temperature. The highest value of A = 2533 MHz was taken at 240 K and A = 1392 MHz at 373 K.

Ferromagnetic interactions were observed in samples N111, N112, N150, and N151. The greatest effect was observed in PbTe (N111). The temperature of the substrate has a large impact on the ferromagnetic properties; the lower it is, the better the ferromagnetic properties, while the admixture of indium gives similar properties at a higher temperature of the substrate. The temperature range of 200–240 K was also defined, in which it was possible to obtain DMS.

An interesting effect was observed for the temperature around 200–240 K. There is a change in the slope of the temperature dependencies g_{eff}, especially for the ferromagnetic line $g_{eff} \approx 5$ (Figure 6), and it is in correlation with the electrical measurements (Figure 11). A likely source of ferromagnetism is localized electrons that, under increasing temperature, move into the conduction band and undergo delocalization. This is particularly clear for sample N111, for which the Curie temperature is near this range, and for this sample, we observe a decay of ferromagnetic properties above 240 K. Since the Curie temperatures of the other samples are low, this effect is not observed, and we only see changes in the value of the spin–orbit coupling for the g-factor example. Further research of these materials is planned to improve the ferromagnetic properties and obtain the highest Curie temperature.

Author Contributions: Conceptualization, A.W. and I.S.; methodology, A.W.; investigation, A.W., B.C. and I.V.; data curation, A.W. and B.C.; writing—original draft preparation, A.W. and B.C.; writing—review and editing, A.W., B.C., I.S., I.V. and R.Ś.; supervision, I.S. and R.Ś. All authors have read and agreed to the published version of the manuscript.

Funding: This research received no external funding.

Institutional Review Board Statement: Not applicable.

Informed Consent Statement: Not applicable.

Data Availability Statement: The data presented in this study are available on request from the corresponding author.

Conflicts of Interest: The authors declare no conflict of interest.

Sample Availability: Samples of the compounds are not available from the authors.

References

1. Dornhaus, R.; Nimtz, G.; Schlicht, B. *Narrow-Gap Semiconductors*; Springer: Berlin/Heidelberg, Germany, 1983; Volume 98. [CrossRef]
2. Bauer, G.; Pascher, H.; Zawadzki, W. Magneto-optical properties of semimagnetic lead chalcogenides. *Semicond. Sci. Technol.* **1992**, *7*, 703–723. [CrossRef]
3. Theocharous, E. Absolute linearity measurements on a PbSe detector in the infrared. *Infrared Phys. Technol.* **2007**, *50*, 63–69. [CrossRef]
4. Isber, S.; Charar, S.; Gratens, X.; Fau, C.; Averous, M. EPR study of the hyperfine structure of ion iSSe, anTe single crystals. *Phys. Rev. B Condens. Matter Mater. Phys.* **1996**, *54*, 7634–7636. [CrossRef]
5. Dietl, T. Origin of ferromagnetic response in diluted magnetic semiconductors and oxides. *J. Phys. Condens. Matter* **2007**, *19*, 165204. [CrossRef]
6. D'Souza, R.; Cao, J.; Querales-Flores, J.D.; Fahy, S.; Savić, I. Electron-phonon scattering and thermoelectric transport in p -type PbTe from first principles. *Phys. Rev. B* **2020**, *102*, 115204. [CrossRef]
7. Lach-Hab, M.; Keegan, M.; Papaconstantopoulos, D.A.; Mehl, M.J. Electronic structure calculations of PbTe. *J. Phys. Chem. Solids* **2000**, *61*, 1639–1645. [CrossRef]
8. Murphy, R.M.; Murray, É.D.; Fahy, S.; Savić, I. Broadband phonon scattering in PbTe-based materials driven near ferroelectric phase transition by strain or alloying. *Phys. Rev. B* **2016**, *93*, 104304. [CrossRef]
9. Dietl, T. Spintronics and ferromagnetism in wide-band-gap semiconductors. *AIP Conf. Proc.* **2005**, *772*, 56–64.
10. Liu, C.; Yun, F.; Morkoç, H. Ferromagnetism of ZnO and GaN: A Review. *J. Mater. Sci. Mater. Electron.* **2005**, *16*, 555–597. [CrossRef]
11. Dietl, T. Origin of ferromagnetism and nano-scale phase separations in diluted magnetic semiconductors. *Phys. E Low-Dimens. Syst. Nanostruct.* **2006**, *35*, 293–299. [CrossRef]
12. Story, T.; Wilamowski, Z.; Grodzicka, E.; Witkowska, B.; Dobrowolski, W. Electron Paramagnetic Resonance of Cr in PbTe. *Acta Phys. Pol. A* **1993**, *84*, 773–775. [CrossRef]
13. Łazarczyk, P.; Story, T.; Arciszewska, M.; Gałęzka, R.R. Magnetic phase diagram of Pb1-x-ySnyMnxTe semimagnetic semiconductors. *J. Magn. Magn. Mater.* **1997**, *169*, 151–158. [CrossRef]
14. Story, T.; Swüste, C.H.W.; Eggenkamp, P.J.T.; Swagten, H.J.M.; de Jonge, W.J.M. Electron paramagnetic resonance knight shift in semimagnetic (diluted magnetic) semiconductors. *Phys. Rev. Lett.* **1996**, *77*, 2802–2805. [CrossRef]
15. Troncoso, J.F.; Aguado-Puente, P.; Kohanoff, J. Effect of intrinsic defects on the thermal conductivity of PbTe from classical molecular dynamics simulations. *J. Phys. Condens. Matter* **2020**, *32*, 045701. [CrossRef]
16. Dobrowolski, W.; Kossut, J.; Story, T. II-VI and IV-VI Diluted Magnetic Semiconductors-New Bulk Materials and Low-Dimensional Quantum Structures. *Handb. Magn. Mater.* **2003**, *15*, 289–377.
17. Streetman, B.G.; Banerjee, S.K. *Solid State Electronic Devices*; Prentice Hall: Hoboken, NJ, USA, 2016; Volume 24, ISBN 0-13-025538-6.
18. Bukała, M.; Sankowski, P.; Buczko, R.; Kacman, P. Crystal and electronic structure of PbTe/CdTe nanostructures. *Nanoscale Res. Lett.* **2011**, *6*, 126. [CrossRef]
19. Boukhris, N.; Meradji, H.; Korba, S.A.; Drablia, S.; Ghemid, S.; Hassan, F.E.H. First principles calculations of structural, electronic and thermal properties of lead chalcogenides PbS, PbSe and PbTe compounds. *Bull. Mater. Sci.* **2014**, *37*, 1159–1166. [CrossRef]
20. Ivanova, Y.A.; Ivanou, D.K.; Streltsov, E.A. Electrochemical deposition of PbTe onto n-Si(1 0 0) wafers. *Electrochem. Commun.* **2007**, *9*, 599–604. [CrossRef]
21. Loke, G.; Yan, W.; Khudiyev, T.; Noel, G.; Fink, Y. Recent Progress and Perspectives of Thermally Drawn Multimaterial Fiber Electronics. *Adv. Mater.* **2020**, *32*, 1904911. [CrossRef]
22. Yan, W.; Dong, C.; Xiang, Y.; Jiang, S.; Leber, A.; Loke, G.; Xu, W.; Hou, C.; Zhou, S.; Chen, M.; et al. Thermally drawn advanced functional fibers: New frontier of flexible electronics. *Mater. Today* **2020**, *35*, 168–194. [CrossRef]
23. Xin, J.; Basit, A.; Li, S.; Danto, S.; Tjin, S.C.; Wei, L. Inorganic thermoelectric fibers: A review of materials, fabrication methods, and applications. *Sensors* **2021**, *21*, 3437. [CrossRef] [PubMed]
24. Bartkowski, M.; Northcott, D.J.; Reddoch, A.H. Superhyperfine structure in the EPR spectra of Mn^{2+} ions in PbTe. *Phys. Rev. B* **1986**, *34*, 6506–6508. [CrossRef] [PubMed]
25. Grossfeld, T.; Sheskin, A.; Gelbstein, Y.; Amouyal, Y. Microstructure evolution of Ag-alloyed PbTe-based compounds and implications for thermoelectric performance. *Crystals* **2017**, *7*, 281. [CrossRef]
26. Minikayev, R.; Safari, F.; Katrusiak, A.; Szuszkiewicz, W.; Szczerbakow, A.; Bell, A.; Dynowska, E.; Paszkowicz, W. Thermostructural and elastic properties of PbTe and Pb0.884Cd0.116Te: A combined low-temperature and high-pressure x-ray diffraction study of Cd-substitution effects. *Crystals* **2021**, *11*, 1063. [CrossRef]
27. Ginting, D.; Lin, C.C.; Rhyee, J.S. Synergetic approach for superior thermoelectric performance in PbTe-PbSe-PbS quaternary alloys and composites. *Energies* **2019**, *13*, 72. [CrossRef]

28. Ben-Ayoun, D.; Sadia, Y.; Gelbstein, Y. Compatibility between co-metallized PbTe thermoelectric legs and an Ag-Cu-In brazing alloy. *Materials* **2018**, *11*, 99. [CrossRef]
29. Gainza, J.; Serrano-Sánchez, F.; Biskup, N.; Nemes, N.M.; Martínez, J.L.; Fernández-Díaz, M.T.; Alonso, J.A. Influence of nanostructuration on PbTe alloys synthesized by arc-melting. *Materials* **2019**, *12*, 3783. [CrossRef]
30. Sato, K.; Katayama-Yoshida, H. Ab initio study on the magnetism in ZnO-, ZnS-, ZnSe- and ZnTe-based diluted magnetic semiconductors. *Phys. Status Solidi Basic Res.* **2002**, *229*, 673–680. [CrossRef]
31. Stefaniuk, I.; Obermayr, W.; Popovych, V.D.; Cieniek, B.; Rogalska, I. EPR Spectra of Sintered Cd $1-x$ Cr x Te Powdered Crystals with Various Cr Content. *Materials* **2021**, *14*, 3449. [CrossRef]
32. Hu, L.; Huang, J.; He, H.; Zhu, L.; Liu, S.; Jin, Y.; Sun, L.; Ye, Z. Dual-donor (Zni and VO) mediated ferromagnetism in copper-doped ZnO micron-scale polycrystalline films: A thermally driven defect modulation process. *Nanoscale* **2013**, *5*, 3918–3930. [CrossRef]
33. Zhuo, S.Y.; Liu, X.C.; Xiong, Z.; Yang, J.H.; Shi, E.W. Ionized zinc vacancy mediated ferromagnetism in copper doped ZnO thin films. *AIP Adv.* **2012**, *2*, 012184. [CrossRef]
34. Hu, L.; Zhu, L.P.; He, H.P.; Ye, Z.Z. Optical demagnetization in defect-mediated ferromagnetic ZnO:Cu films. *Appl. Phys. Lett.* **2014**, *104*, 062405. [CrossRef]
35. Hu, L.; Zhu, L.; He, H.; Ye, Z. Unexpected magnetization enhancement in hydrogen plasma treated ferromagnetic (Zn,Cu)O film. *Appl. Phys. Lett.* **2014**, *105*, 072414. [CrossRef]
36. Hu, L.; Cao, L.; Li, L.; Duan, J.; Liao, X.; Long, F.; Zhou, J.; Xiao, Y.; Zeng, Y.J.; Zhou, S. Two-dimensional magneto-photoconductivity in non-van der Waals manganese selenide. *Mater. Horiz.* **2021**, *8*, 1286–1296. [CrossRef] [PubMed]
37. Kanchana, S.; Jay Chithra, M.; Ernest, S.; Pushpanathan, K. Violet emission from Fe doped ZnO nanoparticles synthesized by precipitation method. *J. Lumin.* **2016**, *176*, 6–14. [CrossRef]
38. Bogomolova, L.D.; Jachkin, V.A.; Prushinsky, S.A.; Dmitriev, S.A.; Stefanovsky, S.V.; Teplyakov, Y.G.; Caccavale, F. Paramagnetic species induced by ion implantation of Pb^+ and C^+ ions in oxide glasses. *J. Non. Cryst. Solids* **1998**, *241*, 174–183. [CrossRef]
39. Zayachuk, D.M.; Ilyina, O.S.; Mikityuk, V.I.; Shlemkevych, V.V.; Kaczorowski, D. Unusual paramagnetic centers in PbTe undoped crystals. *Solid State Sci.* **2014**, *38*, 30–34. [CrossRef]
40. Aminov, L.K.; Zverev, D.G.; Mamin, G.V.; Nikitin, S.I.; Silkin, N.I.; Yusupov, R.V.; Shakhov, A.A. EPR of Pb^{3+} ion in $LiBaF_3$ crystals. *Appl. Magn. Reson.* **2006**, *30*, 175–184. [CrossRef]
41. Solntsev, V.P.; Mashkovtsev, R.I.; Davydov, A.V.; Tsvetkov, E.G. EPR study of coordination of Ag and Pb cations in BaB_2O_4 crystals and barium borate glasses. *Phys. Chem. Miner.* **2008**, *35*, 311–320. [CrossRef]
42. Gawlińska-Nęcek, K.; Wlazło, M.; Socha, R.; Stefaniuk, I.; Major, Ł.; Panek, P. Influence of conditioning temperature on defects in the double Al_2O_3/ZnO layer deposited by the ald method. *Materials* **2021**, *14*, 1038. [CrossRef]
43. Srinivasu, V.V.; Lofland, S.E.; Bhagat, S.M.; Ghosh, K.; Tyagi, S.D. Temperature and field dependence of microwave losses in manganite powders. *J. Appl. Phys.* **1999**, *86*, 1067–1072. [CrossRef]
44. Alvarez, G.; Zamorano, R. Characteristics of the magnetosensitive non-resonant power absorption of microwave by magnetic materials. *J. Alloys Comp.* **2004**, *369*, 231–234. [CrossRef]
45. Montiel, H.; Alvarez, G.; Betancourt, I.; Zamorano, R.; Valenzuela, R. Correlations between low-field microwave absorption and magnetoimpedance in Co-based amorphous ribbons. *Appl. Phys. Lett.* **2005**, *86*, 072503. [CrossRef]
46. Ray, A.; Pillai, P.S.; Krupashankara, M.S.; Satyanarayana, B.S. Nanomechanical Properties of Aluminium Thin Films on Polycarbonate Substrates Using Nanoindentation. *NanoTrends A J. Nanotechnol. Appl.* **2006**, *17*, 23–30.

Article

Spin-Label Electron Paramagnetic Resonance Spectroscopy Reveals Effects of Wastewater Filter Membrane Coated with Titanium Dioxide Nanoparticles on Bovine Serum Albumin

Krisztina Sebők-Nagy [1], Zoltán Kóta [1], András Kincses [1], Ákos Ferenc Fazekas [2], András Dér [1], Zsuzsanna László [2] and Tibor Páli [1,*]

[1] Institute of Biophysics, Biological Research Centre Szeged, 6726 Szeged, Hungary; seboknagy.krisztina@brc.hu (K.S.-N.); kota.zoltan@brc.hu (Z.K.); kincses.andras@brc.hu (A.K.); der.andras@brc.hu (A.D.)

[2] Department of Biosystems Engineering, Faculty of Engineering, University of Szeged, 6725 Szeged, Hungary; fazekas@mk.u-szeged.hu (Á.F.F.); zsizsu@mk.u-szeged.hu (Z.L.)

* Correspondence: pali.tibor@brc.hu

Citation: Sebők-Nagy, K.; Kóta, Z.; Kincses, A.; Fazekas, Á.F.; Dér, A.; László, Z.; Páli, T. Spin-Label Electron Paramagnetic Resonance Spectroscopy Reveals Effects of Wastewater Filter Membrane Coated with Titanium Dioxide Nanoparticles on Bovine Serum Albumin. *Molecules* **2023**, *28*, 6750. https://doi.org/10.3390/molecules28196750

Academic Editor: Yordanka Karakirova

Received: 11 August 2023
Revised: 19 September 2023
Accepted: 20 September 2023
Published: 22 September 2023

Copyright: © 2023 by the authors. Licensee MDPI, Basel, Switzerland. This article is an open access article distributed under the terms and conditions of the Creative Commons Attribution (CC BY) license (https://creativecommons.org/licenses/by/4.0/).

Abstract: The accumulation of proteins in filter membranes limits the efficiency of filtering technologies for cleaning wastewater. Efforts are ongoing to coat commercial filters with different materials (such as titanium dioxide, TiO_2) to reduce the fouling of the membrane. Beyond monitoring the desired effect of the retention of biomolecules, it is necessary to understand what the biophysical changes are in water-soluble proteins caused by their interaction with the new coated filter membranes, an aspect that has received little attention so far. Using spin-label electron paramagnetic resonance (EPR), aided with native fluorescence spectroscopy and dynamic light scattering (DLS), here, we report the changes in the structure and dynamics of bovine serum albumin (BSA) exposed to TiO_2 (P25) nanoparticles or passing through commercial polyvinylidene fluoride (PVDF) membranes coated with the same nanoparticles. We have found that the filtering process and prolonged exposure to TiO_2 nanoparticles had significant effects on different regions of BSA, and denaturation of the protein was not observed, neither with the TiO_2 nanoparticles nor when passing through the TiO_2-coated filter membranes.

Keywords: wastewater cleaning; polyvinylidene fluoride filter membrane; titanium dioxide nanoparticles; P25; serum albumin; spin-label EPR; fluorescence; dynamic light scattering

1. Introduction

The efficiency of the filtering of biomolecules certainly depends on the effect of the surface of the filter on the biomolecules. The separation of proteins from aqueous solutions can be efficiently achieved by membrane filtration. However, membrane fouling seriously limits its application when proteins tend to adsorb to the surface of the membrane pores, hence reducing the pass-through efficiency and, ultimately, the lifespan of the membranes. An ideal filter membrane would keep proteins in the pre-filter space without losing its filtering efficiency. Until now, several new methods have been developed to produce antifouling membranes, such as membrane modification by nanoparticles (e.g., titanium dioxide, TiO_2). Although a number of modified membranes were developed, there is only limited information available regarding whether the applied nanoparticles cause any changes in the protein structure, altering their filterability and other biophysical properties [1–3]. Our previous results showed that coating membranes with TiO_2 kept the membrane structure intact; however, it surprisingly worsened the retention of bovine serum albumin (BSA) [4,5]. This phenomenon can only be partially explained by changes in the roughness and morphology of the membrane surface, and previous investigations have also raised the possibility of changes in the protein structure as well [6]. In the present study, we address precisely this aspect of filtering out proteins from wastewater. We have chosen BSA as the model protein in this work, not only because other teams [7–11] and we also have used it in our recent relevant studies [4,5,12], but also because it is a well-known protein, with plenty of biophysical data about its water-soluble state [13–18], that can serve as reference for comparison. Spin-label EPR spectroscopy is our main technique here because it has proven to be among the most powerful techniques in studying serum albumins [19–22]. Commercial PVDF filter membranes were modified by coating them with TiO_2 nanoparticles (P25), and the effects of the filtration through them and the contact with TiO_2 on the protein structure were investigated by spin-label EPR and native fluorescence spectroscopy, as well as DLS, to reveal the potential changes in the protein structure.

Apart from their diverse physiological functions, serum albumins are the major vehicles of fatty acid transport through blood plasma, as they can bind several long-chain fatty acids with high affinity [23–26]. Due to this fatty-acid-binding function of BSA [27–31] and because spin-label EPR spectroscopy played a crucial role in identifying and characterising the fatty-acid-binding sites of serum albumins [27,30–33], we have used a spin-labelled stearic acid analogue (5-SASL) to detect changes in the fatty acid binding of the protein in its process-relevant interactions with the membrane and the nanoparticles. Fatty acid binding to serum albumins has proven to be sensitive to the physical state of the protein (affected by, e.g., pH, ligand-induced allosteric modulation, and temperature) [25,27,28,32]. We have also used two common maleimide-type spin labels binding to unblocked Cys residues to detect changes in the dynamics of the spin-labelled Cysteines and structural changes in their vicinity [21,34–37]. According to the crystal structure [38], the labelled residue is likely to be Cys34 (in Figure 1, the blue-coloured residue in the sequence and structure [39,40]) because the others are participating in disulfide bridges [21,28].

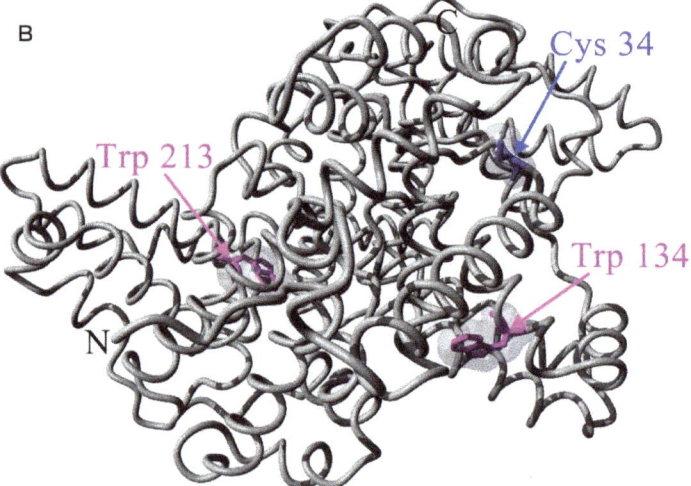

Figure 1. The amino acid sequence (**A**) and an illustrative 3-dimensional structure (**B**) of BSA (PDB i.d. 4s5f). The autofluorescence residues (W, Trp) and the spin-label binding site (C, Cys) are highlighted and coloured in magenta and blue, respectively.

Since BSA has fluorescent residues, it can be studied with fluorescence spectroscopy without attaching a fluorescent dye to the protein [41–43]. Indeed, the two Tryptophan residues of BSA (in Figure 1, magenta-coloured) are sensitive to conformational changes via the altered polarity and rotational dynamics [42–45]. Therefore, we have carried out fluorescence spectroscopic measurements for BSA in the absence and presence of TiO_2 nanoparticles and for filtrated BSA. Since BSA may be present in different states in an aqueous solution depending on its concentration and other conditions, such as folded or denatured monomeric or multimeric, aggregated and micellar forms [10,14,22,46], we tested changes in the particle size distribution of BSA upon the above different treatments using dynamic light scattering (DLS).

2. Results and Discussion

2.1. Chemical Oxygen Demand (COD)

Ultrafiltration experiments were carried out to investigate the effect of a TiO_2 coating on the filtration performance of composite membranes, while microfiltration was used in another series of experiments aimed at investigating the potential effect of TiO_2 nanoparticles on BSA structure and filtration behaviour. In this case, the aim of microfiltration was to separate the TiO_2 nanoparticles from the BSA solution. The BSA rejection of the pristine and modified membranes in the presence (BSA@PVDF/TiO_2) or absence of TiO_2 (BSA@PVDF) is illustrated in Table 1.

Table 1. BSA retention during filtration of BSA through PVDF or PVDF/TiO_2 composite membrane.

Sample	[BSA]/M	Membrane	R%
BSA	1.5×10^{-5}	PVDF	92
BSA	1.5×10^{-5}	PVDF/TiO_2	28
BSA(TiO_2)	1.5×10^{-5}	PVDF	91
BSA(TiO_2)	4.9×10^{-6}	PVDF	37
BSA	4.9×10^{-6}	PVDF	37

It was found that the pristine membrane rejects more BSA than the modified one. This is a very surprising result; thus, we also checked if the higher BSA permeability of the membrane was caused by the contact of the BSA with TiO_2 nanoparticles. The BSA was mixed with TiO_2; then, the BSA was separated from the TiO_2 by microfiltration, and the clean BSA solution was filtrated through the PVDF membrane (BSA(TiO_2)@PVDF). It was found that the contact of the BSA with TiO_2 resulted in no noticeable change in the filtration performance after it was separated from the TiO_2. The filtration alone through cellulose acetate (CA) membrane caused an approx. 5% BSA loss from the 1.5×10^{-5} M solution, while the retention during the filtration through the PVDF membrane was dependent on the BSA concentration: in three-fold-diluted solutions, the retention decreases. Nevertheless, this cannot explain the decreased retention of the BSA filtrated through the TiO_2-modified membrane; thus, in further experiments, the potential effect of the contact between the TiO_2 and BSA on BSA structure was investigated.

2.2. Dynamic Light Scattering (DLS)

We tested the particle size distribution of the BSA in our samples with dynamic light scattering (DLS). Table 2 summarises our data and the literature data, along with the different states of BSA in an aqueous environment. The DLS curves for the stock (6×10^{-5} M) and the 100-fold-diluted BSA solution can be seen in Figure 2.

A narrow peak at around a mean particle size of 1.7 nm, and two very broad and highly overlapping peaks at 24.4 and 85–102 or 105–200 nm are present in the BSA stock solutions (Figure 2A). The 100-fold-diluted BSA solution (Figure 2B) results in the disappearance of the peaks at 1.7 nm, and the appearance of a new peak at ~11 nm can be observed, along with the strong reduction in the broad (20–1000 nm) scattering region. There are a wide range of particle sizes reported in the literature for BSA in different concentration regions and conditions as summarised in Table 2.

Table 2. Particle size of BSA.

[BSA]/10⁻⁶ M	d1/nm	d2/nm	Particle Type	Medium	Reference
0.6		11	dimers (d2)	H_2O	own result
0.65		8.9	compact aggregates (d2)	pH 7.2, 10 mM phosphate buffer	[16]
4.5	10		monomer (d1)	pH 7.0, H_2O	[17]
5	7.3	13.5	monomer (d1), dimer (d2)	pH 7.4, H_2O	[46]
10	3.4		monomer (d1)	pH 7.0, H_2O	[18]
25		10.6	compact aggregates (d2)	pH 7.2, 10 mM phosphate buffer	[16]
60	1.7	24–200	unfolded monomers (d1) large aggregates (d2)	H_2O	own result
100	5.2		undefined particle (d1)	pH 7.4, H_2O	[47]
120.4	2	22	denaturated monomers (d1), aggregation of denaturated monomers (d2)	H_2O	[48]
120.4		12.4	dimer (d2)	pH 7.4, 0.01 M PBS buffer	[48]

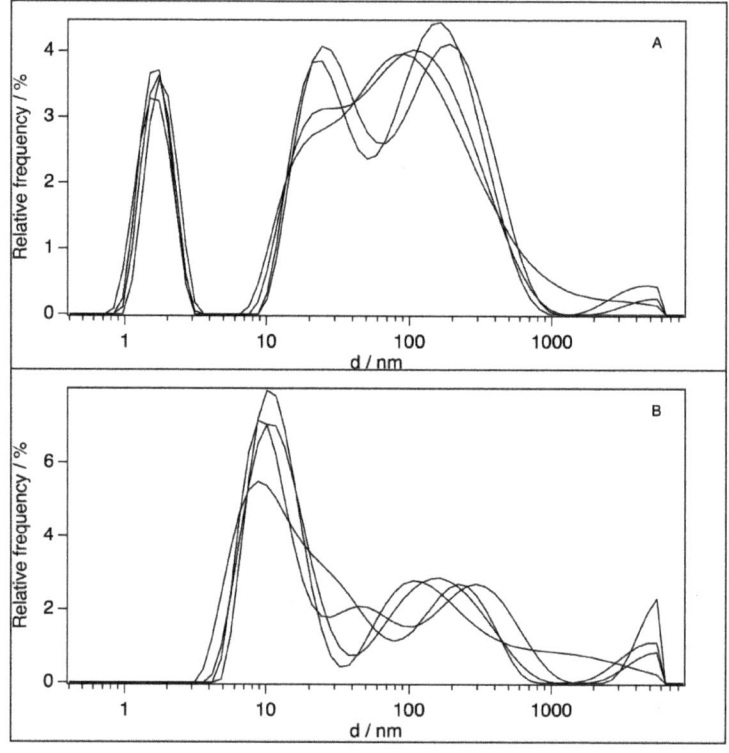

Figure 2. Particle size distribution in BSA solutions in water: [BSA] = 6×10^{-5} M (**A**), and [BSA] = 6×10^{-7} M (**B**). The different curves represent independent experiments.

Comparing our results with this set of previously reported data, we can conclude that the 11 nm peak in our 100-fold-diluted solution is likely to correspond to dimers, whereas the non-diluted BSA dispersion contains large aggregates and a very minor contribution of small (probably unfolded) monomers (considering the logarithmic scale of the x-axis) [48]. It is important to note that, in the present work, the concentration of BSA was 10 or 100 times higher (for the diluted and the stock solutions, respectively) than the 6.5–6.9×10^{-7} M critical micelle concentration (CMC) [16]. Therefore, the disappearance or dissolution of micelles upon their filtering and interaction with TiO_2 can be excluded. It should be noted here that the average size of the membrane pore was estimated to be ~14 nm by Howe and Clark [49], whereas the size of the primer TiO_2 particles is ~25.4 nm, but they form aggregates in water with a diameter of ~1 μm [50]. There is, therefore, a very low probability for the TiO_2 particles to pass through the membrane pores; consequently, the TiO_2 content must be very low in the filtrated BSA sample.

2.3. Samples for Fluorescence and EPR Spectroscopy

Regarding the industrial process of wastewater filtration, we have used four types of samples in our spectroscopic measurements: (i) 6×10^{-5} M BSA in water, (ii) 2×10^{-2} M TiO_2 in water, (iii) BSA-TiO_2 mixture in water where [BSA] = 6×10^{-5} M and [TiO_2] = 2×10^{-2} M, and (iv) 6×10^{-5} M BSA filtrated through composite ultrafilter membranes. These samples are marked as "BSA", "TiO_2", "BSA+TiO_2", and "filtrated BSA", respectively, in the EPR and fluorescence spectroscopic experiments. Some samples had to be diluted 30-fold for optimal conditions in the fluorescence experiments, as mentioned below. Other than this, we did not apply any centrifugation, washing, filtering, or other separation techniques on the samples, to exclude any dilution and concentration effects. Therefore, the free (unbound) spin probe was not removed from the EPR samples. It is important to note that, due to the filtration, the protein concentration in the filtrated BSA samples was only 1.8×10^{-5} M—that is, 3.3-fold lower than in the BSA and BSA+TiO_2 samples. Since BSA is sensitive to pH [27,43], which might be affected by a high concentration of TiO_2, we have measured the pH of the above four types of samples and obtained values of 7.4, 4.9, 7.0, and 7.3 for the BSA, TiO_2, BSA+TiO_2, and filtrated BSA samples, respectively. (We did not buffer the pH to the same value in the different samples because it would be incompatible with the industrial wastewater filtration process.) The corresponding pH values for the 30-fold-diluted samples were 6.7, 6.1, 6.7, and 6.4, respectively. This result shows that the pH effect of the TiO_2 is compensated for by the ~2.5-times-higher concentration of BSA.

2.4. Fluorescence Spectroscopy

BSA has two Tryptophan residues (Trp134 and Trp213; see Figure 1) with fluorescence emission maxima at ~348 nm in pure water [51]. The fluorescence from the Tyrosines is negligible compared to that of the Tryptophanes [42]. We have made fluorescence spectroscopic measurements on all four types of samples from their stock solutions. Table 3 contains the wavelengths of the emission maxima (λ_{max}) (from Gaussian fits) of the BSA-containing samples and the (weighted) mean fluorescence emission maximum ($<\lambda_F>$) as calculated by Equation (1) [52].

$$<\lambda_F> = \frac{\sum f_\lambda \cdot \lambda}{\sum f_\lambda} \quad (1)$$

where λ is the wavelength and f_λ is the emission intensity at λ.

The emission maximum (λ_{max}) and the mean emission maximum ($<\lambda_F>$) for the filtrated sample were 340.2 nm and 356.9 nm, respectively, which are red-shifted by 1.3 nm and 4.6 nm, respectively, relative to the BSA sample (with λ_{max} = 338.9 and $<\lambda_F>$ = 352.3 nm). The BSA+TiO_2 sample could not be measured at this concentration because of strong light scattering. Therefore, the samples were diluted 30-fold to reduce the disturbing light scattering (Figure 3). Dilution alone resulted in a negligible blue shift (0.5 nm) for the BSA and red shift (0.5 nm) for the filtrated BSA samples relative to the original samples in the case of the fitted emission maximum. The mean emission maximum showed little blue shift (1.9 nm) for the filtrated BSA. The diluted BSA+TiO$_2$ sample yielded an intensity maximum at 355.6 nm ($<\lambda_F>$ = 363.3 nm), which means a strong red shift relative to the BSA and filtrated BSA (Figure 3, Table 3). It should be noted that TiO$_2$ did not give a measurable contribution to the fluorescence spectra (Figure 3).

Table 3. Tryptophan emission maxima.

Sample	[BSA]/M	Fitted Emission Maximum (λ_{max})/nm	Mean Emission Maximum ($<\lambda_F>$)/nm
BSA	6×10^{-5}	338.9	352.3
Filtrated BSA	1.8×10^{-5}	340.2	356.9
BSA+TiO$_2$	6×10^{-5}	-	-
BSA	2×10^{-6}	338.4	352.3
Filtrated BSA	6×10^{-7}	340.7	355.0
BSA+TiO$_2$	2×10^{-6}	355.6	363.3

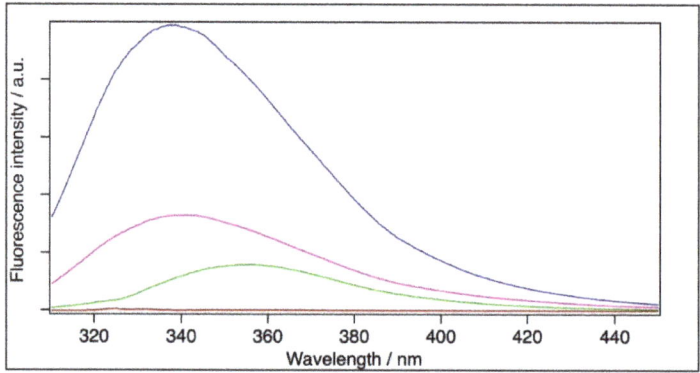

Figure 3. Representative fluorescence spectra. Blue: BSA ([BSA] = 2×10^{-6} M), magenta: filtrated BSA ([BSA] = 6×10^{-7} M), green: BSA+TiO$_2$ ([BSA] = 2×10^{-6} M, [TiO$_2$] = 6.7×10^{-4} M), and red: TiO$_2$ ([TiO$_2$] = 6.7×10^{-4} M).

In the case of Tryptophan, any negative charge in the environment of the pyrrole ring or any positive charge close to the benzene ring causes a change in the electron densities of both rings [44], hence resulting in a bathochromic shift in the fluorescence spectrum. Based on quantum mechanics (DFT) calculations on TiO$_2$, it has been reported that the charge state of Ti is +3 and the oxygen is −1.5 in the molecule [53]. This means that TiO$_2$ is very polar and able to induce a change in electron density both in the benzene and pyrrol rings. It should be noted that, whereas the red shift caused by the filtration is modest and comparable to the dilution effects, that caused by a direct interaction with TiO$_2$ is much larger. The reduction in intensity for the filtrated BSA is simply a dilution effect, since the protein concentration is ~3.3-fold lower in that sample than in the BSA (control) and BSA+TiO$_2$ samples. However, it is striking that the fluorescence intensity of the BSA+TiO$_2$ sample is almost seven-fold smaller than that of BSA in water at the same concentration (Figure 3). This loss of intensity can be explained by fluorescence quenching, since it has also been reported that, apart from causing a red shift, TiO$_2$ nanoparticles can quench fluorescence upon the formation of a higher-order complex [54]. It can, therefore, be assumed that the observed large red shift and loss of intensity in the spectrum of the BSA+TiO$_2$ sample is caused by the TiO$_2$ molecules reaching both the pyrrol and benzene rings of the Trp residues. In contrast to the BSA+TiO$_2$ sample, the relatively short interaction of the BSA with TiO$_2$ during filtration results in no red shift and no fluorescence quenching. This means that the TiO$_2$–BSA interaction during filtration is either too short to be effective on the Trp residues, and/or the effect is reversible. The relatively small difference between the emission maxima of the BSA and filtrated sample can be explained by a small amount of TiO$_2$ getting in the waste during filtration.

2.5. EPR Spectroscopy

The shape of the continuous-wave (CW) EPR spectrum of the spin labels attached to biological molecules is directly sensitive to the speed, amplitude, and symmetry of the rotational dynamics of the label constrained by the orienting potential of its environment [55–58]. We have utilised both known types of labelling targets in BSA: fatty acid bound to the protein [27] and Cys residues with a free sulfhydryl group [21,37]. We have used 5-SASL as a spin-labelled fatty acid analogue, and 5-MSL and MTSL as covalent labels of the Cys residues. These spin labels are shown in Figure 4.

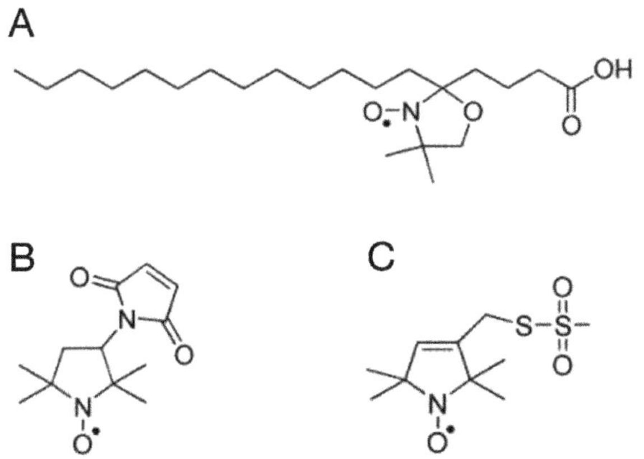

Figure 4. The chemical structure of the spin labels used for the EPR measurements. The EPR spectrum is originating from the unpaired electron located in a π^* orbital of the N-O bound: (**A**) 5-SASL, (**B**) 5-MSL, and (**C**) MTSL.

All three labels have some solubility in water, and, since unbound labels were not removed from the EPR samples, we recorded their spectra also in the absence of BSA and TiO$_2$ in order to identify the different spectral components in the protein-labelled samples. It should be noted that, since TiO$_2$ is diamagnetic, it did not contribute at all to the EPR spectra of the spin labels, as expected.

2.5.1. Spin Labelling with 5-SASL

Figure 5 shows the CW EPR spectra of 5-SASL in water and in the three different aqueous BSA-containing samples, i.e., in the BSA, BSA+TiO$_2$, and filtrated BSA.

The spectra are normalised to the same integrated intensity (second integral), so they represent the same number of spins. The 5-SASL in water has a single component EPR spectrum of three sharp lines, as expected for a spin label freely rotating in a solvent (see, e.g., [30]). The protein-containing samples are qualitatively similar to those in previous EPR reports on spin-labelled fatty acids in the presence of serum albumins [20,27,28,30,33], showing composite spectra with different contributions from a mobile and an immobile component. Since the mobile component was identical with that of 5-SASL in water, optimised subtraction allowed us to determine the shape and relative contribution of the spectrum from 5-SASL bound to BSA, and also the mobile component (see [59] for a detailed description of the technique). The spectra of BSA and BSA+TiO$_2$ have almost only the immobile component, with only a few percentages of the mobile component. In the spectrum of the filtrated BSA, the mobile component is apparently the dominant one (providing ~2/3 of the integrated intensity). However, it should be kept in mind that the BSA concentration in that sample is 3.3-fold lower than in the other two BSA-containing sample types. The relative contributions of the mobile and immobile components are given in Table 4.

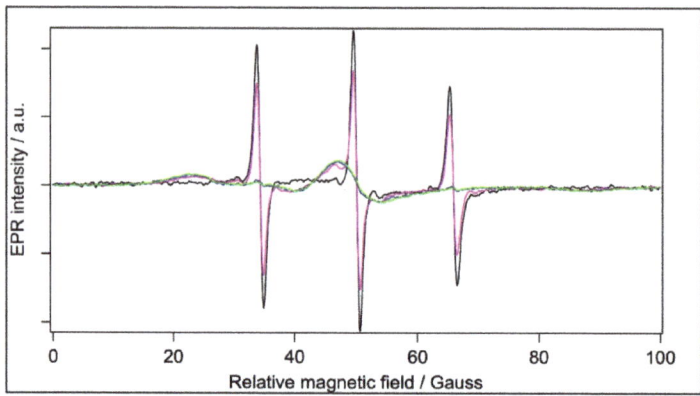

Figure 5. Normalised CW EPR spectra of samples labelled with 5-SASL. Black: water, blue: BSA in water, magenta: filtrated BSA, and green: BSA+TiO$_2$. [5-SASL] = 1.2 × 10^{-4} M.

Table 4. EPR parameters of samples spin-labelled with 5-SASL. [5-SASL] = 1.2 × 10^{-4} M.

Sample	Mobile Component/%	Immobile Component/%	τ/ns	2A$_{zz}$/G
Water	100	-	0.1206	-
BSA	2.4	97.6	-	64.62
Filtrated BSA	71	29	0.1312	62.3–64.1
BSA+TiO$_2$	1.0	99.0	-	65.05

Considering the lower protein concentration in the filtrated BSA sample, we can conclude that all BSA-containing samples bind the same (high) amount of 5-SASL per protein. With the exception of the filtrated BSA sample, the 5-SASL spectra are dominated by either the mobile or the immobile component to an extent that the minor component is too weak for meaningful component separation. The mobile and immobile spectral shapes require different spectrum analyses in order to obtain data on the rotational dynamics of the doxyl group of 5-SASL: The sharp hyperfine lines of the mobile component (the dominating spectra of the BSA in water and filtrated BSA) can be used to derive the mean rotation correlation time using the Kivelson formula (Equation (2)) [60,61].

$$\tau_R(\text{ns}) = 0.65 \cdot W_0(\text{Gauss}) \cdot \left(\sqrt{\frac{h_0}{h_{-1}}} - \sqrt{\frac{h_0}{h_{+1}}} \right) \tag{2}$$

where τ_R is the rotation correlation time in ns, W_0 is the line width of the narrowest (central) peak, and h_{+1}, h_0, and h_{-1} are the intensities of the hyperfine lines. On the other hand, if the rotational dynamics are slow (on the EPR time window) or limited in amplitude, then the EPR spectrum shows an anisotropic spread between the line positions at the minimum and maximum hyperfine-splitting values, corresponding to the z-axis of the doxyl group being oriented perpendicular or parallel, respectively, to the magnetic field. In our experiments, the immobile components represent rotational dynamics constrained by the local environment of the spin label. In this case, the orientational order determines the inner and outer hyperfine splitting (A$_{zz}$), from which the orientational order parameter can usually be determined (see, e.g., [61,62]). However, the immobile components do not sufficiently expose the inner splittings and only the outer splitting constant can be easily determined (which is half of the magnetic field difference between the first local maximum and last local minimum); they are still in a monotonic relationship with the order parameter, with a larger outer splitting meaning a higher order. The rotation correlation times and outer hyperfine-splitting constants are also reported in Table 4. The rotation is very fast for water and for the filtrated BSA, and the values (0.12062 ns and 0.13124 ns, respectively) are very close to each other, meaning that the mobile component in the filtrated BSA spectrum corresponds to the unbound free 5-SASL. The presence of a strong immobile component in the BSA-containing samples suggests that the fatty-acid-binding sites are preserved both in the presence of TiO_2 and after filtration. Similar outer-splitting values (64.6 G and 65.1 G) were obtained for BSA and BSA+TiO_2. There is some uncertainty with regard to the outer splitting for the filtrated BSA because of the big signal/noise ratio of the decomposed spectrum, but it is safe to say it is ~2 G smaller for this sample. This means more disorder for the 5-SASL bound to BSA in the presence of TiO_2, which may be an indication to loosen the fatty-acid-binding pocket of the BSA by TiO_2.

2.5.2. Spin Labelling with 5-MSL

BSA contains 35 Cystein residues. However, according to the experimental structure of the protein (illustrated in Figure 1) and the literature data, only one (Cys34) is not involved in forming S-S bridges, with a non-bonded sulfhydryl group offering a single unique site for covalent labelling with maleimide-type spin labels [28,63–66]. The EPR spectra of 5-MSL added to water and the different BSA samples are shown in Figure 6A.

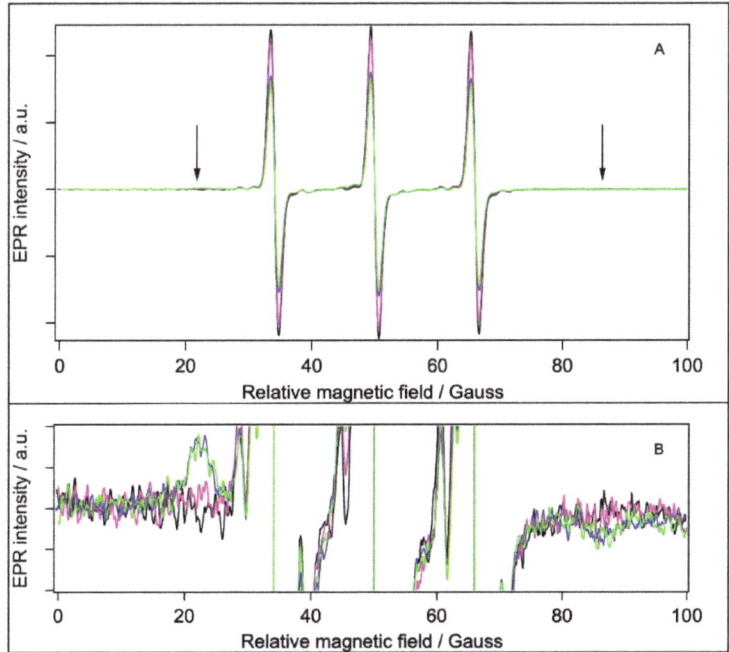

Figure 6. Normalised CW EPR spectra of samples labelled with 5-MSL. Black: water, blue: BSA, magenta: filtrated BSA, and green: BSA+TiO_2. (**A**) Full spectra; (**B**) 73-fold-magnified spectrum to visualise the immobile component (shown with arrows on panel A). [5-MSL] = 1.2×10^{-4} M.

According to its chemical structure (Figure 4B), the covalent binding of 5-MSL to a Cystein yields a relatively rigid connection between the N-O bound (bearing the unpaired electron) and the protein backbone; hence, an immobile component is expected for a covalently bound 5-MSL [21,22,28,34,67]. The EPR spectra in Figure 6A have very similar shapes but different intensity. (It should be kept in mind that the spectra represent the same number of spins; therefore, smaller amplitudes indicate line broadening.) However, there is a clear sign of a weak immobile component present in the spectra of the BSA and BSA+TiO$_2$ samples (and, to lesser extent, in the filtrated BSA, which, however, has a 3.3-fold lower protein concentration), as indicated by the arrows and in the bottom part of the figure. We could separate the mobile and immobile components and analyse them in a similar way as with the 5-SASL spectra, and the extracted parameters are presented in Table 5.

Table 5. EPR parameters of samples spin-labelled with 5-MSL. [5-MSL] = 1.2×10^{-4} M.

Sample	Mobile Component/%	Immobile Component/%	τ/ns	$2A_{zz}$/G
Water	100	-	0.0043	-
BSA	71	29	0.0082	63.92
Filtrated BSA	93	7	0.0056	cc. 61 G *
BSA+TiO$_2$	67	33	0.0084	64.20

* The immobile component of this sample is very noisy (because of its small contribution to the composite spectrum); hence, the outer splitting is much less certain than for the other samples.

Comparing the corresponding immobile fractions for 5-SASL (Table 4) and 5-MSL (Table 5), it is evident that the labelling of BSA is more efficient with 5-SASL than with 5-MSL. The immobile fraction, the rotational correlation time, and the outer splitting values are very close to those of the BSA and BSA+TiO$_2$ samples, whereas the filtrated BSA has an apparently much smaller immobile component (but with a smaller outer splitting, similarly to 5-SASL). However, if we correct for the 3.3-fold lower protein concentration in the filtrated BSA than in the other two BSA-containing samples, the immobile component per protein is the same. It should be noted that the rotational correlation times are at the fast limit of the EPR time window for spin labels [55,68]. These values most probably correspond to the free rotation of the spin label in water (5-MSL is much smaller than 5-SASL). Again, the reduced fraction of the immobile component in the filtrated sample is mostly caused by the similarly lower concentration of BSA in the filtrated sample than in the BSA and BSA+TiO$_2$ samples. The variation in τ_R is likely to be caused by the colliding of the unbound label with the surface of the slow-tumbling BSA (reducing the mean τ_R), which agrees with the observation that τ_R is smaller in the filtrated BSA sample, in which the BSA concentration is smaller than in the other two BSA-containing samples. A similar explanation applies when comparing the 5-MSL in water vs. in the BSA samples. It should be also noted that the outer splitting (hence, the orientational disorder) is comparable to those obtained with 5-SASL.

2.5.3. Spin Labelling with MTSL

Spin-labelling with MTSL was performed in the same way as with 5-MSL. MTSL (Figure 4C) has a disulfide bridge between the doxyl ring and the free sulfhydryl group of the Cysteine amino acid, which provides more rotational freedom and flexibility to this label compared with 5-MSL. It is, therefore, expected that MTSL displays higher mobility than 5-MSL. The EPR spectra of MTSL in water and the BSA-containing samples are shown in Figure 7.

Through optimised spectral subtractions, we found that the MTSL spectra have three components: in addition to the expected mobile and an immobile component, we observed a five-peak component known to originate from the dimeric (biradical) form of this type of label [69]. Fortunately, the five peaks of the dimeric form do not overlap with the three lines of the immobile component of the normal, monomeric form. Therefore, by using diluted samples to change the contribution of the five-peak component (not shown), the three components could be separated: We first obtained the pure five-peak component from the series of spectra of MTLS at different label concentrations. Then, the five-peak component was subtracted from the composite spectra, leaving the normal mobile plus immobile components, which were then treated with the routine technique for the two components [59]. Although we see the immobile component as observed earlier [31], it is almost undetectable; hence, the outer splittings could not be determined. (Again, it should be kept in mind that the filtrated BSA sample has a 3.3-fold lower protein concentration than the other two BSA-containing samples.) Table 6 reports the fractional contribution of the mobile and immobile component and the rotational correlation time derived from the mobile component of the monomeric MTSL in water and in the BSA-containing samples.

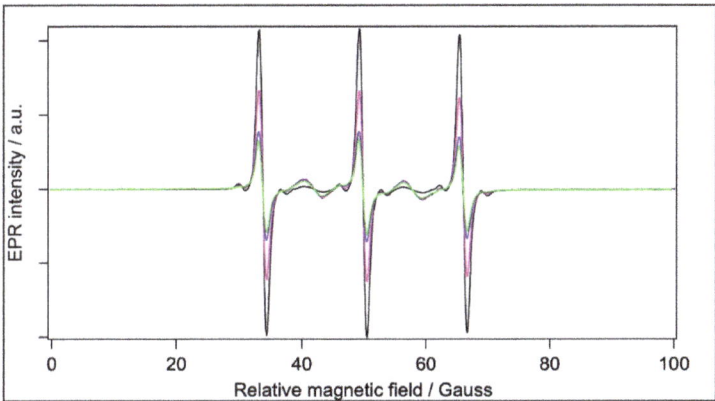

Figure 7. Normalised CW EPR spectra of samples labelled with MTSL. Black: water, blue: BSA, magenta: filtrated BSA, and green: BSA+TiO$_2$, [MTSL] = 1.2 × 10^{-4} M.

Table 6. EPR parameters of samples spin-labelled with MTSL. [MTSL] = 1.2 × 10^{-4} M.

Sample	Mobile Component/%	Immobile Component/%	τ/ns	2A$_{zz}$/G
Water	100	-	0.0094	-
BSA	97–98	2–3	0.0303	-
Filtrated BSA	100	-	0.0224	-
BSA+TiO$_2$	97–98	2–3	0.0536	-

According to its mobile component, the rotation of the monomeric MTSL is ~5–6 times slower in the BSA-containing samples than in water, and ~5–8 times slower than the 5-MSL in the corresponding samples. We cannot exclude the idea that the mobile components are from bound MTSL, but it is more likely that the mobile component is from MTSL monomers in water, and the effective rotational correlation time is increased in BSA-containing samples by diffusional collisions with the protein micelles, which is confirmed by the apparent concentration dependence of τ_R in the dilution experiment (not shown). As opposed to the results with 5-MSL (where the BSA+TiO$_2$ and BSA spectra report similar rotational dynamics, as shown in Table 5), the MTSL spectrum of the BSA+TiO$_2$ yields somewhat slower rotational correlation time than those of BSA or filtrated BSA (Table 6). This is probably due to the higher accessibility of TiO$_2$ to bound MTSL because of its longer linker, or to its preferable interaction in water.

3. Materials and Methods

3.1. Chemicals and Sample Preparation

Bovine serum albumin (BSA) and TiO$_2$ Aeroxide P25 were of analytical grade and purchased from VWR International (Debrecen, Hungary). Pristine and modified Polyvinylidene fluoride (PVDF) with molecular weight cut-off (MWCO) values of 100 kDa and ultrafilter (UF) membranes Kynar 400 PVDF with 100 kDa molecular weight cut-off (MWCO) were purchased from New Logic Research Inc. (Minden, NV, USA). The spin labels, 5-doxyl stearic acid (5-SASL) and 3-maleimido-proxyl (5-MSL), were purchased from Sigma (Budapest, Hungary), and 1-(oxyl-2,2,5,5-tetramethylpyrrolidine-3-methyl) methanethiosulfonate (MTSL) was obtained from Santa Cruz Biotechnology (Dallas, TX, USA). Ethanol was obtained from Molar Chemicals (Halásztelek, Hungary).

The aim of ultrafiltration experiments was to investigate the effect of TiO_2 coating on filtration performance of composite membranes. The ultrafilter membranes were prepared by coating PVDF with inorganic TiO_2 nanoparticles. Commercial PVDF membranes were used to prepare nanoparticle-coated membranes by a physical deposition method [4]. For this purpose, 0.04 g of commercial TiO_2 was added to 100 mL of ultrapure water and ultrasonicated for 3 min. Then, the ultrasonicated suspension was filtrated through a membrane in a dead-end filtration device (Millipore, XFUF04701, Merck KGa, Darmstadt, Germany) at 0.3 MPa and dried for 1 h at room temperature before use. The procedure of the filtration has been performed as described previously [4]. Briefly, BSA rejection tests were performed at 0.1 MPa. In each filtration, 250 mL of water or model solution was filtrated until a 200 mL of permeate was obtained. Concentration of BSA was measured before and after filtration by measuring chemical oxygen demand (COD) of the solution, and BSA rejection was calculated. To check the potential effect of TiO_2 nanoparticles on BSA filtration performance, another sample was prepared as follows: 0.04 g TiO_2 and 250 mg BSA were dissolved in 250 mL ultrapure water and stirred (350 rpm) for 1.5 h, then filtrated through a 0.22 μm cellulose acetate (CA) syringe filter to separate BSA from TiO_2. The clear BSA solution was then filtrated through the unmodified PVDF ultrafilter membrane (labelled as BSA/TiO_2@PVDF), and COD rejection was calculated. Rejection was calculated using the following equation:

$$R(\%) = \frac{c_0 - c}{c_0} \times 100 \tag{3}$$

where c_0 and c are the concentrations of feed and permeate solutions, respectively.

In order to examine the potential effect of TiO_2 on BSA, four types of samples were analysed with EPR and fluorescent spectroscopy: (1) BSA 6×10^{-5} M in water, (2) TiO_2 2×10^{-2} M in water, (3) BSA-TiO_2 mixture in water with [BSA] = 6×10^{-5} M and [TiO_2] = 2×10^{-2} M in water, and (4) BSA 6×10^{-5} M filtrated through composite ultrafilter membranes. The bandgap of TiO_2 P25 is around 3.2 eV, which means that 385 nm UV light can activate it, but it has poor activity under visible light [70]. The effect of light exposure was checked and excluded during experiments; nevertheless, the filtration experiments were performed in dark conditions.

The spin labels were dissolved in ethanol (6×10^{-3} M) for the stock solutions. Protein concentration was checked by Lowry method [71]. The BSA concentration was always reduced by the filtration; concentrations for filtrated BSA stocks were determined to be typically ~1.8×10^{-5} M. However, the protein concentration in the (unfiltrated) BSA and BSA+TiO_2 samples were not adjusted to that of filtrated ones because we wanted to avoid any change in the original samples.

3.2. Experimental Procedures and Data Analysis

COD was measured by the standard potassium dichromate oxidation method using standard test tubes (Lovibond, Tintometer Gmbh, Dortmund, Germany) and digestion and COD measurements were carried out in a COD digester (Lovibond, ET 108, Tintometer Gmbh, Dortmund, Germany) and a COD photometer (Lovibond PC-CheckIt, Tintometer Gmbh, Dortmund, Germany).

Dynamic light scattering measurements for BSA were made by a Zetasizer Nano ZS (Malvern Panalytical, Malvern, UK) instrument. The principle of the DLS technique is that fine particles and molecules that are in Brownian motion diffuse at a speed relative to their size. To measure the diffusion speed, the scattered light of a He-Ne laser (633 nm) illuminating the particles is recorded. The light intensity at a specific angle fluctuates with time, which is recorded using a sensitive avalanche photodiode detector. The autocorrelation function of this time-series curve is informative in terms of the size distribution of the hydrodynamic radius of the particles [72]. The samples contained 0.03, 0.3, 0.6, and 6×10^{-5} M BSA in water.

Fluorescence measurements were carried out by a Fluorolog-3 (FL3-222) modular spectrofluorimeter (Horiba (Jobin Yvon), Kyoto, Japan) with an excitation wavelength of 295 nm. Stocks and 30-times-diluted solutions of BSA, BSA+TiO$_2$, and filtrated BSA were used for the measurements.

Continuous-wave EPR measurements were carried out on a Bruker ELEXSYS-II E580 X-band spectrometer at room temperature, with the following instrument settings: scan range 3300–3400 G; microwave frequency ~9.42 GHz; microwave power 9.464 mW; modulation frequency 100 kHz; modulation amplitude 0.4 G; resolution of field axes 1024; to scans 16; sweep time 40.96 s. The first series of experiments had the molar ratio of protein and spin labels 1:2 where the concentration of the spin label was 1.2×10^{-4} M with the volume ratio of 2%. Further sample series were 5- or 10-fold-diluted either for both the stocks and the spin labels, hence preserving all the molar ratios, or only for spin labels. There was no column chromatography or other separation technique used to remove unbound spin labels to avoid any further agitation of the samples.

Data analysis and fitting the plotting were performed with *Igor Pro*, version 8.02 (Wave Metrics, Lake Oswego, OR, USA). Molecular graphics were made with *YASARA*, version 21.12.19 [73].

4. Conclusions

Our previous results on the effect of coating filter membranes with TiO$_2$ nanoparticles on the membrane and filtration [4,5] could not explain the changes in the BSA structure observed here. Although filtering causes a 3.3-fold reduction in the BSA concentration, it is still significantly above the CMC. Therefore, the observed effects, as reported by EPR and fluorescence spectroscopy, are not caused by changes in the particle sizes. We can also exclude an indirect structural effect of TiO$_2$ on BSA through the pH, because the observed pH changes in the BSA+TiO$_2$ and filtrated BSA relative to the BSA samples are too small (<0.4 units) to be effective [43]. Unfolded protein monomers represent a negligible phase of BSA as opposed to the large micelles or assemblies present in all the samples over the wide range of size distribution. These large particles represent low mobility in the EPR time scale, which is reflected by the immobile component of the EPR spectra of the tightly bound fraction of the spin labels (5-SASL and 5-MSL). The potentially more loosely bound MTSL has a negligible bound fraction. It is important to note that the source of the fluorescence signal and the EPR of spin-labelled fatty acids are not as specific as the EPR of the uniquely unblocked Cys34, since there are two Trp residues and up to seven fatty-acid-binding pockets in BSA. Indeed, there are indications that all the binding sites in an albumin may have bound fatty acid even at the lowest levels of ligand loading [31]. In addition, Cys34 (in domain I) is relatively close to the protein surface [65,66], and, when spin-labelled, it has been reported to be more accessible to collisional interaction with a water-soluble paramagnetic relaxant than the acyl chain of non-covalently bound spin-labelled fatty acids (located in hydrophobic channels in the protein) [28]. Here, we found that the Trp regions are more sensitive to the presence of TiO$_2$ nanoparticles, whereas both the stearic-acid-binding pockets and the environment of the spin-labelled Cysteine are more sensitive to the mechanical stress on BSA caused by the filtration process: the labels are more mobile in these samples, despite a comparatively very similar level of binding. The difference between these processes (as compared to the BSA dissolved in water) is that filtration exerts a mechanical stress on the protein and its aggregated forms (micelles) with a relatively short duration of contact with the TiO$_2$. On the contrary, mixing BSA with TiO$_2$ does not result in a mechanical stress to the BSA, but the protein is permanently exposed to interaction with TiO$_2$. The visual appearance of the BSA solutions suggests that there is no precipitation of

BSA. In addition, our spectroscopic data show that BSA has a native-like fluorescence after filtration and retains the control level of the binding of fatty acid and 5-MSL spin label both after filtration and in the presence of TiO_2, evidencing that the structural integrity of the BSA is preserved in the present experiments.

It can be concluded that the filtration process and prolonged exposure to TiO_2 have a significant effect on different regions of BSA. However, the unfolding, denaturing, or precipitation of the protein by the prolonged presence of TiO_2 or filtration through a TiO_2-coated filter membrane was not observed. Obviously, further research is needed, for instance, in the direction of slower filtering and/or different coating materials with higher elimination efficiency, filtration, and TiO_2 effect on the particle size of BSA and its binding of fatty acids and CD spectroscopy of changes in the secondary structure.

Author Contributions: K.S.-N. made the EPR experiments and helped with interpreting the data and drafting the manuscript. Z.K. made the fluorescence experiments. A.K. made and A.D. evaluated and interpreted the DLS experiments. Á.F.F. made the COD experiments. Z.L. proposed the project, supervised the COD experiments, provided the membrane filters and TiO_2 nanoparticles, and contributed to the interpretation of the results. T.P. co-ordinated the work, designed the spectroscopy experiments and interpreted the results, and provided the concept of the manuscript and refined it to its final form. All authors have read and agreed to the published version of the manuscript.

Funding: This research was funded in part by the National Research, Development, and Innovation Office of Hungary—PD-143268.

Institutional Review Board Statement: Not applicable.

Informed Consent Statement: Not applicable.

Data Availability Statement: All experimental data presented in this study are available upon request from the corresponding author.

Acknowledgments: We would like to acknowledge the support provided by the National Research, Development, and Innovation Office of Hungary—PD-143268.

Conflicts of Interest: The authors declare no conflict of interest. There were no external influences on the authors in the design of the study; in the collection, analyses, or interpretation of the data; in the writing of the manuscript; or in the decision to publish the results.

Sample Availability: Not available.

References

1. Wangoo, N.; Suri, C.R.; Shekhawat, G. Interaction of Gold Nanoparticles with Protein: A Spectroscopic Study to Monitor Protein Conformational Changes. *Appl. Phys. Lett.* **2008**, *92*, 133104. [CrossRef]
2. Gheshlaghi, Z.N.; Riazi, G.H.; Ahmadian, S.; Ghafari, M.; Mahinpour, R. Toxicity and Interaction of Titanium Dioxide Nanoparticles with Microtubule Protein. *Acta Biochim. Biophys. Sin.* **2008**, *40*, 777–782. [CrossRef] [PubMed]
3. Bardhan, M.; Mandal, G.; Ganguly, T. Steady State, Time Resolved, and Circular Dichroism Spectroscopic Studies to Reveal the Nature of Interactions of Zinc Oxide Nanoparticles with Transport Protein Bovine Serum Albumin and to Monitor the Possible Protein Conformational Changes. *J. Appl. Phys.* **2009**, *106*, 034701. [CrossRef]
4. Jigar, E.; Bagi, K.; Fazekas, Á.; Kertész, S.; Veréb, G.; László, Z. Filtration of BSA through TiO_2 Photocatalyst Modified PVDF Membranes. *Desalin. Water Treat.* **2020**, *192*, 392–399. [CrossRef]
5. Sisay, E.J.; Fazekas, Á.F.; Gyulavári, T.; Kopniczky, J.; Hopp, B.; Veréb, G.; László, Z. Investigation of Photocatalytic PVDF Membranes Containing Inorganic Nanoparticles for Model Dairy Wastewater Treatment. *Membranes* **2023**, *13*, 656. [CrossRef]
6. Saptarshi, S.R.; Duschl, A.; Lopata, A.L. Interaction of Nanoparticles with Proteins: Relation to Bio-Reactivity of the Nanoparticle. *J. Nanobiotechnol.* **2013**, *11*, 26. [CrossRef]
7. Cao, X.; Ma, J.; Shi, X.; Ren, Z. Effect of TiO_2 Nanoparticle Size on the Performance of PVDF Membrane. *Appl. Surf. Sci.* **2006**, *253*, 2003–2010. [CrossRef]
8. Liu, C.; Guo, Y.; Hong, Q.; Rao, C.; Zhang, H.; Dong, Y.; Huang, L.; Lu, X.; Bao, N. Bovine Serum Albumin Adsorption in Mesoporous Titanium Dioxide: Pore Size and Pore Chemistry Effect. *Langmuir* **2016**, *32*, 3995–4003. [CrossRef]
9. Wang, X.; Zhou, M.; Meng, X.; Wang, L.; Huang, D. Effect of Protein on PVDF Ultrafiltration Membrane Fouling Behavior under Different pH Conditions: Interface Adhesion Force and XDLVO Theory Analysis. *Front. Environ. Sci. Eng.* **2016**, *10*, 12. [CrossRef]
10. Marquez, A.; Berger, T.; Feinle, A.; Hüsing, N.; Himly, M.; Duschl, A.; Diwald, O. Bovine Serum Albumin Adsorption on TiO_2 Colloids: The Effect of Particle Agglomeration and Surface Composition. *Langmuir* **2017**, *33*, 2551–2558. [CrossRef]

11. Shen, L.; Feng, S.; Li, J.; Chen, J.; Li, F.; Lin, H.; Yu, G. Surface Modification of Polyvinylidene Fluoride (PVDF) Membrane via Radiation Grafting: Novel Mechanisms Underlying the Interesting Enhanced Membrane Performance. *Sci. Rep.* **2017**, *7*, 2721. [CrossRef] [PubMed]
12. Nascimben Santos, E.; Fazekas, Á.; Hodúr, C.; László, Z.; Beszédes, S.; Scheres Firak, D.; Gyulavári, T.; Hernádi, K.; Arthanareeswaran, G.; Veréb, G. Statistical Analysis of Synthesis Parameters to Fabricate PVDF/PVP/TiO_2 Membranes via Phase-Inversion with Enhanced Filtration Performance and Photocatalytic Properties. *Polymers* **2021**, *14*, 113. [CrossRef] [PubMed]
13. Squire, P.G.; Moser, P.; O'Konski, C.T. The Hydrodynamic Properties of Bovine Serum Albumin Monomer and Dimer. *Biochemistry* **1968**, *7*, 4261–4272. [CrossRef] [PubMed]
14. Brahma, A.; Mandal, C.; Bhattacharyya, D. Characterization of a Dimeric Unfolding Intermediate of Bovine Serum Albumin under Mildly Acidic Condition. *Biochim. Biophys. Acta* **2005**, *1751*, 159–169. [CrossRef] [PubMed]
15. Masuelli, M.A. Study of Bovine Serum Albumin Solubility in Aqueous Solutions by Intrinsic Viscosity Measurements. *Adv. Phys. Chem.* **2013**, *2013*, 360239. [CrossRef]
16. Devi, L.B.; Mandal, A.B. BSA can Form Micelle in Aqueous Solution. *J. Surf. Sci. Technol.* **2015**, *31*, 21–29.
17. Li, R.; Wu, Z.; Wangb, Y.; Ding, L.; Wang, Y. Role of pH-Induced Structural Change in Protein Aggregation in Foam Fractionation of Bovine Serum Albumin. *Biotechnol. Rep. (Amst.)* **2016**, *9*, 46–52. [CrossRef]
18. Varga, N.; Hornok, V.; Sebők, D.; Dékány, I. Comprehensive Study on the Structure of the BSA from Extended-to Aged Form in Wide (2-12) pH Range. *Int. J. Biol. Macromol.* **2016**, *88*, 51–58. [CrossRef]
19. de Sousa Neto, D.; Salmon, C.E.; Alonso, A.; Tabak, M. Interaction of Bovine Serum Albumin (BSA) with Ionic Surfactants Evaluated by Electron Paramagnetic Resonance (EPR) Spectroscopy. *Colloids Surf. B Biointerfaces* **2009**, *70*, 147–156. [CrossRef]
20. Neacsu, M.-V.; Matei, I.; Ionita, G. The Extent of Albumin Denaturation Induced by Aliphatic Alcohols: An Epr and Circular Dichroism Study. *Rev. Roum. Chim.* **2017**, *62*, 637–643.
21. Pavićević, A.; Luo, J.; Popović-Bijelić, A.; Mojović, M. Maleimido-proxyl as an EPR Spin Label for the Evaluation of Conformational Changes of Albumin. *Eur. Biophys. J.* **2017**, *46*, 773–787. [CrossRef] [PubMed]
22. Reichenwallner, J.; Oehmichen, M.-T.; Schmelzer, C.; Hauenschild, T.; Kerth, A.; Hinderberger, D. Exploring the pH-Induced Functional Phase Space of Human Serum Albumin by EPR Spectroscopy. *Magnetochemistry* **2018**, *4*, 47. [CrossRef]
23. Spector, A.A.; John, K.; Fletcher, J.E. Binding of Long-chain Fatty Acids to Bovine Serum Albumin. *J. Lipid Res.* **1969**, *10*, 56–67. [CrossRef] [PubMed]
24. Simard, J.R.; Zunszain, P.A.; Hamilton, J.A.; Curry, S. Location of High and Low Affinity Fatty Acid Binding Sites on Human Serum Albumin Revealed by NMR Drug-competition Analysis. *J. Mol. Biol.* **2006**, *361*, 336–351. [CrossRef] [PubMed]
25. Fanali, G.; di Masi, A.; Trezza, V.; Marino, M.; Fasano, M.; Ascenzi, P. Human Serum Albumin: From Bench to Bedside. *Mol. Aspects Med.* **2012**, *33*, 209–290. [CrossRef] [PubMed]
26. Rizzuti, B.; Bartucci, R.; Sportelli, L.; Guzzi, R. Fatty Acid Binding into the Highest Affinity Site of Human Serum Albumin Observed in Molecular Dynamics Simulation. *Arch. Biochem. Biophys.* **2015**, *579*, 18–25. [CrossRef]
27. Ge, M.T.; Rananavare, S.B.; Freed, J.H. ESR Studies of Stearic Acid Binding to Bovine Serum Albumin. *Biochim. Biophys. Acta* **1990**, *1036*, 228–236. [CrossRef]
28. Livshits, V.A.; Marsh, D. Fatty Acid Binding Sites of Serum Albumin Probed by Non-linear Spin-label EPR. *Biochim. Biophys. Acta* **2000**, *1466*, 350–360. [CrossRef]
29. Junk, M.J.; Spiess, H.W.; Hinderberger, D. DEER in Biological Multispin-systems: A Case Study on the Fatty Acid Binding to Human Serum Albumin. *J. Magn. Reson.* **2011**, *210*, 210–217. [CrossRef]
30. Pavićević, A.A.; Popović-Bijelić, A.D.; Mojović, M.D.; Šušnjar, S.V.; Bačić, G.G. Binding of Doxyl Stearic Spin Labels to Human Serum Albumin: An EPR Study. *J. Phys. Chem. B* **2014**, *118*, 10898–10905. [CrossRef]
31. Reichenwallner, J.; Hauenschild, T.; Schmelzer, C.E.H.; Hülsmann, M.; Godt, A.; Hinderberger, D. Fatty Acid Triangulation in Albumins Using a Landmark Spin Label. *Isr. J. Chem.* **2019**, *59*, 1059–1074. [CrossRef]
32. Morrisett, J.D.; Pownall, H.J.; Gotto, A.M. Bovine serum albumin. Study of the Fatty Acid and Steroid Binding Sites using Spin-labeled lipids. *J. Biol. Chem.* **1975**, *250*, 2487–2494. [CrossRef] [PubMed]
33. Perkins, R.C., Jr.; Abumrad, N.; Balasubramanian, K.; Dalton, L.R.; Beth, A.H.; Park, J.H.; Park, C.R. Equilibrium Binding of Spin-labeled Fatty Acids to Bovine Serum Albumin: Suitability as Surrogate Ligands for Natural Fatty Acids. *Biochemistry* **1982**, *21*, 4059–4064. [CrossRef]
34. Benga, G.; Strach, S.J. Interpretation of the Electron Spin Resonance Spectra of Nitroxide-maleimide-labelled Proteins and the Use of this Technique in the Study of Albumin and Biomembranes. *Biochim. Biophys. Acta* **1975**, *400*, 69–79. [CrossRef] [PubMed]
35. Marsh, D.; Livshits, V.A.; Pali, T.; Gaffney, B.J. Recent Development in Biological Spin-label Spectroscopy. In *Spectroscopy of Biological Molecules: New Directions*; Greve, J., Puppels, G.J., Otto, C., Eds.; Kluwer Academic Publishers: Dordrecht, The Netherlands; Boston, MA, USA; London, UK, 1999; pp. 647–650. [CrossRef]
36. Páli, T.; Marsh, D. Structural Studies on Membrane Proteins Using Non-linear Spin Label EPR Spectroscopy. *Cell. Mol. Biol. Lett.* **2002**, *7*, 87–91. [PubMed]
37. Bordignon, E. EPR Spectroscopy of Nitroxide Spin Probes. *eMagRes* **2017**, *6*, 6235–6254. [CrossRef]
38. Bujacz, A. Structures of Bovine, Equine and Leporine Serum Albumin. *Acta Crystallogr. D Biol. Crystallogr.* **2012**, *68*, 1278–1289. [CrossRef]

39. The UniProt Consortium. UniProt: The Universal Protein Knowledgebase in 2023. *Nucleic Acids Res.* **2023**, *51*, D523–D531. [CrossRef]
40. Bujacz, A.; Bujacz, G. Crystal Structure of Bovine Serum Albumin. *Transp. Protein* **2012**. [CrossRef]
41. Celej, M.S.; Montich, G.G.; Fidelio, G.D. Protein Stability Induced by Ligand Binding Correlates with Changes in Protein Flexibility. *Protein Sci.* **2003**, *12*, 1496–1506. [CrossRef]
42. Togashi, D.M.; Ryder, A.G.; Mc Mahon, D.; Dunne, P.; McManus, J. Fluorescence Study of Bovine Serum Albumin and Ti and Sn Oxide Nanoparticles Interactions. In *Diagnostic Optical Spectroscopy in Biomedicine IV, Proceedings of the SPIE-OSA Biomedical Optics, Munich, Germany 17–21 June 2007*; Schweitzer, D., Fitzmaurice, M., Eds.; Optica Publishing Group: Washington, DC, USA, 2007; Volume 6628, p. 6628_61. [CrossRef]
43. Bhattacharya, M.; Jain, N.; Bhasne, K.; Kumari, V.; Mukhopadhyay, S. pH-Induced Conformational Isomerization of Bovine Serum Albumin Studied by Extrinsic and Intrinsic Protein Fluorescence. *J. Fluoresc.* **2011**, *21*, 1083–1090. [CrossRef] [PubMed]
44. Vivian, J.T.; Callis, P.R. Mechanisms of Tryptophan Fluorescence Shifts in Proteins. *Biophys. J.* **2001**, *80*, 2093–2109. [CrossRef] [PubMed]
45. Dos Santos Rodrigues, F.H.; Delgado, G.G.; Santana da Costa, T.; Tasic, L. Applications of Fluorescence Spectroscopy in Protein Conformational Changes and Intermolecular Contacts. *BBA Adv.* **2023**, *3*, 100091. [CrossRef] [PubMed]
46. Li, Y.; Yang, G.; Mei, Z. Spectroscopic and Dynamic Light Scattering Studies of the Interaction between Pterodontic Acid and Bovine Serum Albumin. *Acta Pharm. Sin. B* **2012**, *2*, 53–59. [CrossRef]
47. Polat, H.; Kutluay, G.; Polat, M. Analysis of Dilution Induced Disintegration of Micellar Drug Carriers in the Presence of Inter and Intra Micellar Species. *Colloids Surf. A* **2020**, *601*, 124989. [CrossRef]
48. Burgstaller, C.; Etchart, N.N. Dimerization of Bovine Serum Albumin as Evidenced by Particle Size and Molecular Mass Measurement. Anton Paar GmbH. D51 I A044EN-A. 2018. Available online: https://www.anton-paar.com/corp-en/services-support/document-finder/application-reports/dimerization-of-bovine-serum-albumin-as-evidenced-by-particle-size-and-molecular-mass-measurement/ (accessed on 31 August 2023).
49. Howe, K.J.; Clark, M.M. Fouling of Microfiltration and Ultrafiltration Membranes by Natural Waters. *Environ. Sci. Technol.* **2002**, *36*, 3571–3576. [CrossRef]
50. Bacova, J.; Knotek, P.; Kopecka, K.; Hromádko, L.; Čapek, J.; Nývltová, P.; Bruckova, L.; Schroterova, L.; Sestakova, B.; Palarcik, J.; et al. Evaluating the Use of TiO_2 Nanoparticles for Toxicity Testing in Pulmonary A549 Cells. *Int. J. Nanomed.* **2022**, *17*, 4211–4225. [CrossRef]
51. Teale, F.W.J.; Weber, G. Ultraviolet Fluorescence of the Aromatic Amino Acids. *Biochem. J.* **1957**, *65*, 476–482. [CrossRef]
52. Pocanschi, C.L.; Popot, J.L.; Kleinschmidt, J.H. Folding and Stability of Outer Membrane Protein A (OmpA) from Escherichia Coli in an Amphipathic Polymer, Amphipol A8-35. *Eur. Biophys. J.* **2013**, *42*, 103–118. [CrossRef]
53. Koch, D.; Manzhos, S. On the Charge State of Titanium in Titanium Dioxide. *J. Phys. Chem. Lett.* **2017**, *8*, 1593–1598. [CrossRef]
54. Pandit, S.; Kundu, S. Fluorescence quenching and related interactions among globular proteins (BSA and lysozyme) in presence of titanium dioxide nanoparticles. *Colloids Surf.* **2021**, *628*, 127253. [CrossRef]
55. Freed, J.H. Theory of Slow Tumbling ESR Spectra of Nitroxides. In *Spin Labeling. Theory and Applications*; Berliner, L.J., Ed.; Academic Press: New York, NY, USA, 1976; pp. 53–132.
56. Hubbell, W.L.; Lopez, C.J.; Altenbach, C.; Yang, Z. Technological Advances in Site-directed Spin Labeling of Proteins. *Curr. Opin. Struct. Biol.* **2013**, *23*, 725–733. [CrossRef] [PubMed]
57. Lopez, C.J.; Fleissner, M.R.; Brooks, E.K.; Hubbell, W.L. Stationary-phase EPR for Exploring Protein Structure, Conformation, and Dynamics in Spin-labeled Proteins. *Biochemistry* **2014**, *53*, 7067–7075. [CrossRef] [PubMed]
58. Marsh, D. Spin-label Order Parameter Calibrations for Slow Motion. *Appl. Magn. Reson.* **2018**, *49*, 97–106. [CrossRef] [PubMed]
59. Páli, T.; Kóta, Z. Studying Lipid-Protein Interactions with Electron Paramagnetic Resonance Spectroscopy of Spin-labeled Lipids. *Methods Mol. Biol.* **2019**, *2013*, 529–561. [CrossRef]
60. Kivelson, D. Theory of ESR Linewidths of Free Radicals. *J. Chem. Phys.* **1960**, *33*, 1094–1107. [CrossRef]
61. Páli, T.; Pesti, M. Chapter V: Phase Transition of Membrane Lipids. In *Manual on Membrane Lipids*; Prasad, R., Ed.; Springer: Berlin/Heidelberg, Germany; New York, NY, USA, 1996; pp. 80–111. [CrossRef]
62. Griffith, O.H.; Jost, P.C. Lipid Spin Labels in Biological Membranes. In *Spin Labeling. Theory and Applications*; Berliner, L.J., Ed.; Academic Press: New York, NY, USA, 1976; pp. 453–523. [CrossRef]
63. Peters, T., Jr. Ligand Binding in Albumin. In *All about Albumin*; Academic Press: Cambridge, MA, USA, 1995; pp. 76–132. [CrossRef]
64. Hull, H.H.; Chang, R.; Kaplan, L.J. On the Location of the Sulfhydryl Group in Bovine Plasma Albumin. *Biochim. Biophys. Acta* **1975**, *400*, 132–136. [CrossRef]
65. Sugio, S.; Kashima, A.; Mochizuki, S.; Noda, M.; Kobayashi, K. Crystal Structure of Human Serum Albumin at 2.5 A Resolution. *Protein Eng.* **1999**, *12*, 439–446. [CrossRef]
66. Stewart, A.J.; Blindauer, C.A.; Berezenko, S.; Sleep, D.; Tooth, D.; Sadler, P.J. Role of Tyr84 in Controlling the Reactivity of Cys34 of Human Albumin. *FEBS J.* **2005**, *272*, 353–362. [CrossRef]
67. Griffith, O.H.; McConnell, H.M. A Nitroxide-maleimide Spin Label. *Proc. Natl. Acad. Sci. USA* **1966**, *55*, 8–11. [CrossRef]
68. Lange, A.; Marsh, D.; Wassmer, K.H.; Meier, P.; Kothe, G. Electron Spin Resonance Study of Phospholipid Membranes Employing a Comprehensive Line-Shape Model. *Biochemistry* **1985**, *24*, 4383–4392. [CrossRef] [PubMed]

69. Eaton, S.S.; Woodcock, L.B.; Eaton, G.R. Continuous Wave Electron Paramagnetic Resonance of Nitroxide Biradicals in Fluid Solution. *Concepts Magn. Reson. Part A* **2018**, *47A*, 21426. [CrossRef] [PubMed]
70. Mahy, J.G.; Carcel, C.; Man, M.W.C. Evonik P25 photoactivation in the visible range by surface grafting of modified porphyrins for p-nitrophenol elimination in water. *AIMS Mater.* **2023**, *10*, 437–452. [CrossRef]
71. Lowry, O.H.; Rosebrough, N.J.; Farr, L.; Randall, R.J. Protein Measurement with the Folin Phenol Reagent. *J. Biol. Chem.* **1951**, *193*, 265–275. [CrossRef] [PubMed]
72. Taneva, S.G.; Krumova, S.; Bogár, F.; Kincses, A.; Stoichev, S.; Todinova, S.; Danailova, A.; Horváth, J.; Násztor, Z.; Kelemen, L.; et al. Insights into Graphene Oxide Interaction with Human Serum Albumin in Isolated State and in Blood Plasma. *Int. J. Biol. Macromol.* **2021**, *175*, 19–29. [CrossRef] [PubMed]
73. Krieger, E.; Vriend, G. YASARA View—Molecular Graphics for All Devices—From Smartphones to Workstations. *Bioinform* **2014**, *30*, 2981–2982. [CrossRef]

Disclaimer/Publisher's Note: The statements, opinions and data contained in all publications are solely those of the individual author(s) and contributor(s) and not of MDPI and/or the editor(s). MDPI and/or the editor(s) disclaim responsibility for any injury to people or property resulting from any ideas, methods, instructions or products referred to in the content.

Article

Electron Paramagnetic Resonance Studies of Irradiated Grape Snails (*Helix pomatia*) and Investigation of Biophysical Parameters

Aygun Nasibova [1,2,*], Rovshan Khalilov [1,2], Mahammad Bayramov [1], İslam Mustafayev [1], Aziz Eftekhari [3,4,*], Mirheydar Abbasov [5], Taras Kavetskyy [6,7], Gvozden Rosić [8,*] and Dragica Selakovic [8,*]

1. Institute of Radiation Problems, Ministry of Science and Education Republic of Azerbaijan, AZ1143 Baku, Azerbaijan
2. Department of Biophysics and Biochemistry, Baku State University, AZ1148 Baku, Azerbaijan
3. Department of Biochemistry, Faculty of Science, Ege University, Izmir 35040, Turkey
4. Institute of Molecular Biology & Biotechnologies, Ministry of Science and Education Republic of Azerbaijan, 11 Izzat Nabiyev, AZ1073 Baku, Azerbaijan
5. Institute of Catalysis and Inorganic Chemistry, Ministry of Science and Education Republic of Azerbaijan, AZ1143 Baku, Azerbaijan
6. Department of Biology and Chemistry, Drohobych Ivan Franko State Pedagogical University, 82100 Drohobych, Ukraine
7. Department of Materials Engineering, The John Paul II Catholic University of Lublin, 20-950 Lublin, Poland
8. Department of Physiology, Faculty of Medical Sciences, University of Kragujevac, 34000 Kragujevac, Serbia
* Correspondence: aygunnasibova21@gmail.com (A.N.); ftekhari@ymail.com (A.E.); grosic@medf.kg.ac.rs (G.R.); dragica984@gmail.com (D.S.)

Citation: Nasibova, A.; Khalilov, R.; Bayramov, M.; Mustafayev, İ.; Eftekhari, A.; Abbasov, M.; Kavetskyy, T.; Rosić, G.; Selakovic, D. Electron Paramagnetic Resonance Studies of Irradiated Grape Snails (*Helix pomatia*) and Investigation of Biophysical Parameters. *Molecules* 2023, 28, 1872. https://doi.org/10.3390/molecules28041872

Academic Editor: Yordanka Karakirova

Received: 25 December 2022
Revised: 10 February 2023
Accepted: 10 February 2023
Published: 16 February 2023

Copyright: © 2023 by the authors. Licensee MDPI, Basel, Switzerland. This article is an open access article distributed under the terms and conditions of the Creative Commons Attribution (CC BY) license (https://creativecommons.org/licenses/by/4.0/).

Abstract: A study of grape snails (*Helix pomatia*) using the electron paramagnetic resonance (EPR) spectroscopy method, where shells were exposed to ionizing gamma radiation, indicated that the effect of radiation up to certain doses results in the emergence of magnetic properties in the organism. The identification of the EPR spectra of the body and shell parts of the control and irradiated grape snails separately showed that more iron oxide magnetic nanoparticles are generated in the body part of the grape snail compared to the shells. A linear increase in free radical signals (g = 2.0023) in the body and shell parts of grape snails, and a non-monotonic change in the broad EPR signal (g = 2.32) characterizing iron oxide magnetic nanoparticles was determined depending on the dose of ionizing gamma radiation. Additionally, the obtained results showed that grape snails can be used as bioindicators for examining the ecological state of the environment. At the same time, the radionuclide composition of the body and shell parts of the grape snails and their specific activities were determined by CANBERRA gamma spectroscopy. The FTIR spectra of mucin, a liquid secreted by snails, were recorded.

Keywords: grape snails; stress factors; gamma radiation; magnetic properties; magnetic nanoparticles; EPR signals; radionuclide composition

1. Introduction

To date, some research has been carried out on various living systems that are affected by various stress factors [1–4]. Several studies have investigated the effects of ionizing gamma radiation, ultraviolet rays, drought, humidity, temperature, and other stress factors on living systems [5–7]. Previously, we performed different research on the effect of various stress factors (temperature, gamma radiation, UV radiation, etc.) on plant systems [8–13]. The mechanisms of action of stress factors on various types of tree and shrub plants taken from nature, on seedlings of seeds grown in laboratory conditions, and on chloroplasts isolated from the leaves of higher plants have been studied. In the current study, paramagnetism phenomena occurring in living systems under the influence of stress factors were evaluated.

In our comparative studies, the plants were derived from both ecologically clean and polluted areas, and then, using the EPR method, it was revealed that environmental pollution causes the generation of iron oxide magnetic nanoparticles in plants. The identification of EPR spectra of plants of the same species showed that the intensities of broad EPR signals (g = 2.32, ΔH = 320 G) characterizing iron oxide magnetic nanoparticles in plants growing in ecologically polluted areas are higher than the intensities of the corresponding signals of plants growing in clean areas [14,15]. At the same time, the study of sprouts of some plant seeds (corn (*Zea mays* L.), wheat (*Triticum* L.), and peas (*Cicer arietinum* L.)) irradiated with different doses of ionizing gamma radiation in laboratory conditions showed that the effect of gamma radiation up to certain doses causes the formation of iron oxide magnetic nanoparticles in plants. We have also provided the mechanism of occurrence of this event [10,16,17].

To confirm the results, we continued our research with animal organisms and investigated the effects of radiation factors on laboratory rats *(Wistar albino)* and grape snails with shells *(Helix pomatia)* [18–21].

Some biophysical parameters of grape snails (*Helix pomatia*), which belong to the phylum of mollusks and are commonly found in Absheron (Azerbaijan), were investigated. During the study of the effect of radiation factors on grape snails, it was determined that these factors cause the emergence of magnetic properties in them.

There are several reasons why grape snails with shells are of interest as research objects. First of all, they are distinguished by their high vitality. Thus, snails are very resistant to biological, physical, chemical, and radioactive stress [22–24]. Their blood–vascular systems are open. Snails carry hemocyanin, a protein containing copper molecules, dissolved in blood plasma. Grape snails live in lightly shaded gardens, vineyards, and open areas. Their day is spent hiding in a shell. They feed at night. They mainly eat the green parts of plants [25–27].

In our previous works, we have shown that magnetic nanoparticles, especially magnetite (Fe_3O_4) and maghemite (γ-Fe_2O_3) play an important role in the function of biological systems. These nanoparticles lead to the appearance of magnetic properties in natural systems and the formation of broad EPR signals, which we first discovered in plant leaves [28–32].

For this reason, the study of the mechanisms of formation of nanoparticles of biogenic origin in all living systems is of great interest. With this in mind, in further experiments, we chose grape snails as the research object.

2. Results and Discussion

While conducting research with grape snails, the effect of ionizing gamma radiation on young and old grape snails was studied separately. Note that the age of snails is determined by the size of their shells. Young and old grape snails were placed in plastic containers. Young grape snails were irradiated with doses of 200 Gy, 400 Gy, and 600 Gy, and old grape snails were irradiated with ionizing gamma radiation at doses of 150 Gy, 250 Gy, 400 Gy, 600 Gy, and 800 Gy. Control snails and snails that underwent irradiation with different doses of gamma radiation young and old grape snails were stored in special containers in laboratory conditions at a temperature of 22–25 °C for 60 days. During this period, they were fed the same amount of vegetables (carrots and cucumbers) and water. The grape snails were kept in a regime of 16 h of light and 8 h of darkness.

After 60 days, their shell and body parts were separated, dried at room temperature under natural conditions, and prepared for EPR studies.

Then, EPR spectra of control and irradiated samples were recorded in a wide range of magnetic fields (500–5500 G). It should be noted that consecutive measurements were performed at least five times. Figure 1 shows the EPR spectra of shell parts of the control and irradiated old snails at different doses.

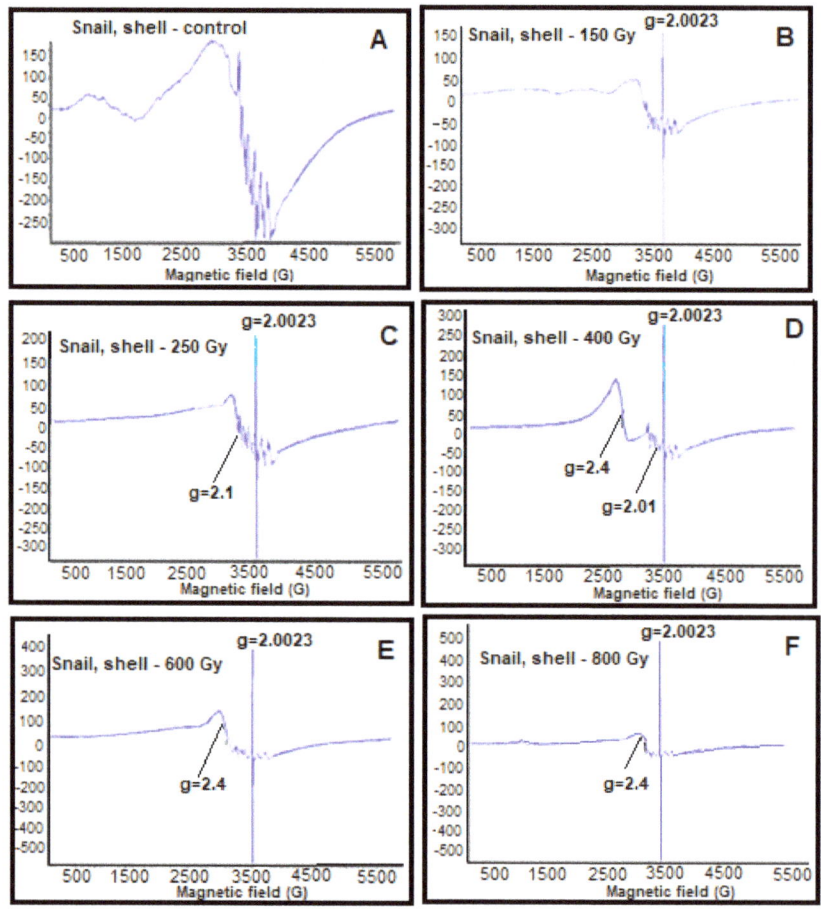

Figure 1. EPR spectra of shell parts of the control and irradiated at different doses of old snails. (**A**) control, (**B**) 150 Gy, (**C**) 250 Gy, (**D**) 400 Gy, (**E**) 600 Gy, and (**F**) 800 Gy.

The parameters of the EPR spectrometer are as follows: center field—3000 G; sweep width—5500 G; resolution—1024 points; frequency—9870 GHz; power—2102 mW; modulation frequency—100 kHz; modulation amplitude—10 G.

The identification of the EPR spectra of the shell parts of old grape snails recorded in Figure 1 illustrates the formation of free radical signals (g = 2.0023), broad EPR signals characterizing iron oxide magnetic nanoparticles (g = 2.4), six-component manganese ion signal with hyperfine structure (g = 2.01), and copper ions signals (g = 2.1). It was also observed that the increase in gamma radiation dose leads to a linear increase in the intensity of free radical signals (g = 2.0023). Depending on the dose of radiation, the monotonically dependent change in the intensity of the free radical signals obtained from the shell parts of grape snails allows us to use them as bioindicator parameters. Therefore, the regular change in the parameters of free radical signals depending on the dose of radiation allows us to use them in the assessment and monitoring of the ecological state of the environment.

Broad EPR signals characteristic of iron oxide magnetic nanoparticles (g = 2.4) appeared in the EPR spectra of the shell parts of grape snails. When the radiation was at 150 Gy, 250 Gy, we observed this signal at a small amplitude. Additionally, when the radiation dose reaches 400 Gy, the intensity of the broad EPR signal characterizing magnetic nanoparticles takes its maximum value. It has been found that irradiation of 600 Gy slightly reduces the intensity of this signal. Irradiation with ionizing gamma radiation at a dose of 800 Gy causes a doubling of the signal intensity.

Thus, as a result of the effect of radiation as a stress factor on snails in certain doses (150 Gy, 250 Gy, and 400 Gy), an increase in free iron ions in their body and the creation of a reducing environment occur. This leads to the formation of nanophase iron oxide particles as a result of biomineralization. However, although the escalation of radiation dose (600 Gy) increases free iron ions in the organism of the grape snails, the weakening of the reducing system decreases the formation of magnetic nanoparticles. The effect of radiation at the highest dose (800 Gy) results in the complete failure of the reducing system, and therefore the formation of nanoparticles does not occur. Thus, the absence of a reducing system leads to the generation of free radicals and reactive oxygen species (ROS). It is known that during stress, as a result of breaks in the living system, the bound iron changes to the free iron form. Due to the Fenton reaction, the increase in iron ions leads to the formation of ROS (for example, hydroxyl radical (HO$^\bullet$); superoxide anion (O2$^{\bullet-}$); hydrogen peroxide (H$_2$O$_2$); hydroxide ion (HO$^-$), etc.). The living system transforms ROS into nano-sized magnetic particles to prevent them from multiplying. Additionally, the shells of snails do not have a reducing system. However, iron-based nanoparticles occur in them. This can be because the shells of grape snails are directly connected with the body and are "fed" through it. Thus, the body parts of grape snails affect the formation of the structure and composition of the shells' parts.

Figure 2 shows the EPR spectra of body parts of the control old grape snails and the old grape snails irradiated at different doses of gamma radiation. As shown, the spectra obtained from the body parts of snails show signals of free radicals (g = 2.0023), broad EPR signals characterizing iron oxide nanoparticles (g = 2.32), and signals of iron ions (g = 3.43). A comparison of the EPR spectra obtained from the shell and body parts of the grape snails shows that the intensity of the generated signals was more intense in the body samples. This may be due to the presence of a reducing system in the body parts of snails. Compared to the control, irradiation up to 250 Gy produces a broad EPR signal with a very high amplitude, which characterizes iron oxide magnetic nanoparticles. Increasing the radiation dose from 400 Gy to 800 Gy leads to a gradual fall in the intensity of this signal.

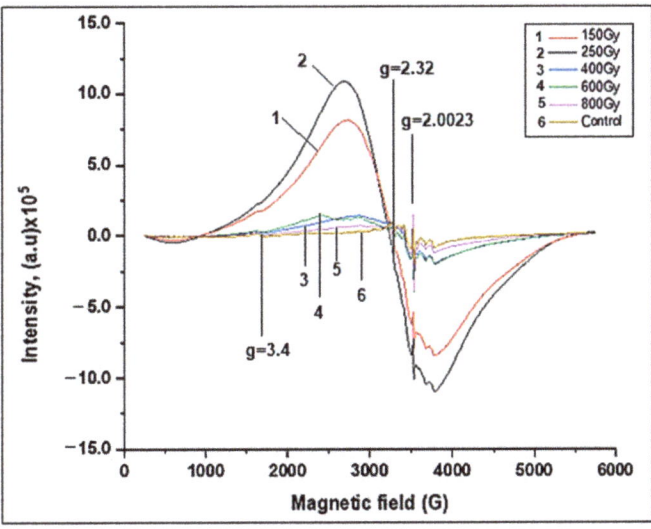

Figure 2. EPR spectra of body parts of old grape snails in the control and irradiated objects.

Certain doses (200 Gy and 350 Gy) of ionizing gamma radiation increase free iron ions in their body, and the presence of a reducing environment leads to the formation of nanophase iron oxide particles inside them as a result of biomineralization. Within the dose increment (600 Gy), again, the weakening of the reducing systems causes the formation of iron oxide magnetic nanoparticles to plummet there. Irradiation at the highest dose of gamma radiation (800 Gy) causes the complete breakdown of the reducing system, so the formation of nanoparticles does not occur. Thus, the absence of a reducing system leads to the formation of free radicals and ROS. This leads to the destruction of the living system.

This means that the formation of iron oxide magnetic nanoparticles as a result of stress has the purpose of self-defense of the organism.

EPR spectra of the body and shell parts of young grape snails were also recorded (Figures 3 and 4). It was identified that the changes in the behavior of the EPR signals shell and body parts of old snails were also observed in the EPR spectra of young grape snails.

However, depending on the dose of gamma radiation in old grape snails, the dynamics of the changes in free radical signals ($g = 2.0023$) and characteristic broad EPR signals of iron oxide magnetic nanoparticles ($g = 2.32$) obtained from body parts were more intense than the dynamics of corresponding signals. This can be explained by the fact that old grape snails have lived longer in nature and have been more exposed to various environmental factors (for example, temperature, humidity, drought, UV radiation, and others).

The recorded spectra of the alterations of the intensity of the free radical signals and the intensity of the EPR signals of iron oxide magnetic nanoparticles depending on the radiation dose are shown graphically (Figures 5 and 6). The obtained result is very important in monitoring and evaluating the ecological condition of the environment. We can say that the regular change in parameters of wide EPR signals characterizing iron oxide magnetic nanoparticles ($g = 2.4$) obtained from the body and shell parts of grape snails under the influence of stress factors is informative in assessing the degree of environmental pollution.

When studying the effect of various stress factors on living systems, it was found that stress factors cause the emergence of new magnetic properties in animal organisms as well as in plant systems.

Thus, during the study of the effect of various stress factors on living systems (plants, animal organisms, and chloroplasts isolated from higher plant leaves) using the EPR method, it was found that any stress factor in certain doses as a result of biomineralization causes the formation of iron oxide magnetic nanoparticles (magnetite—Fe_3O_4 and maghemite—γ—Fe_2O_3) in a living system. These nanoparticles lead to the formation of new magnetic properties in living systems. This result is very promising and relevant in terms of assessment and biomonitoring of the ecological state of the environment. Because our research showed that the parameters of EPR spectra of control and irradiated samples can be used as bioindicator parameters, these signals are informative in environmental monitoring.

In addition, the obtained results are also considered very important in terms of biomedical applications. Thus, our experiments showed that as a result of stress factors up to certain doses, nano-sized iron oxide particles are formed in living systems. These nanoparticles play an important role in medical treatment and diagnostics.

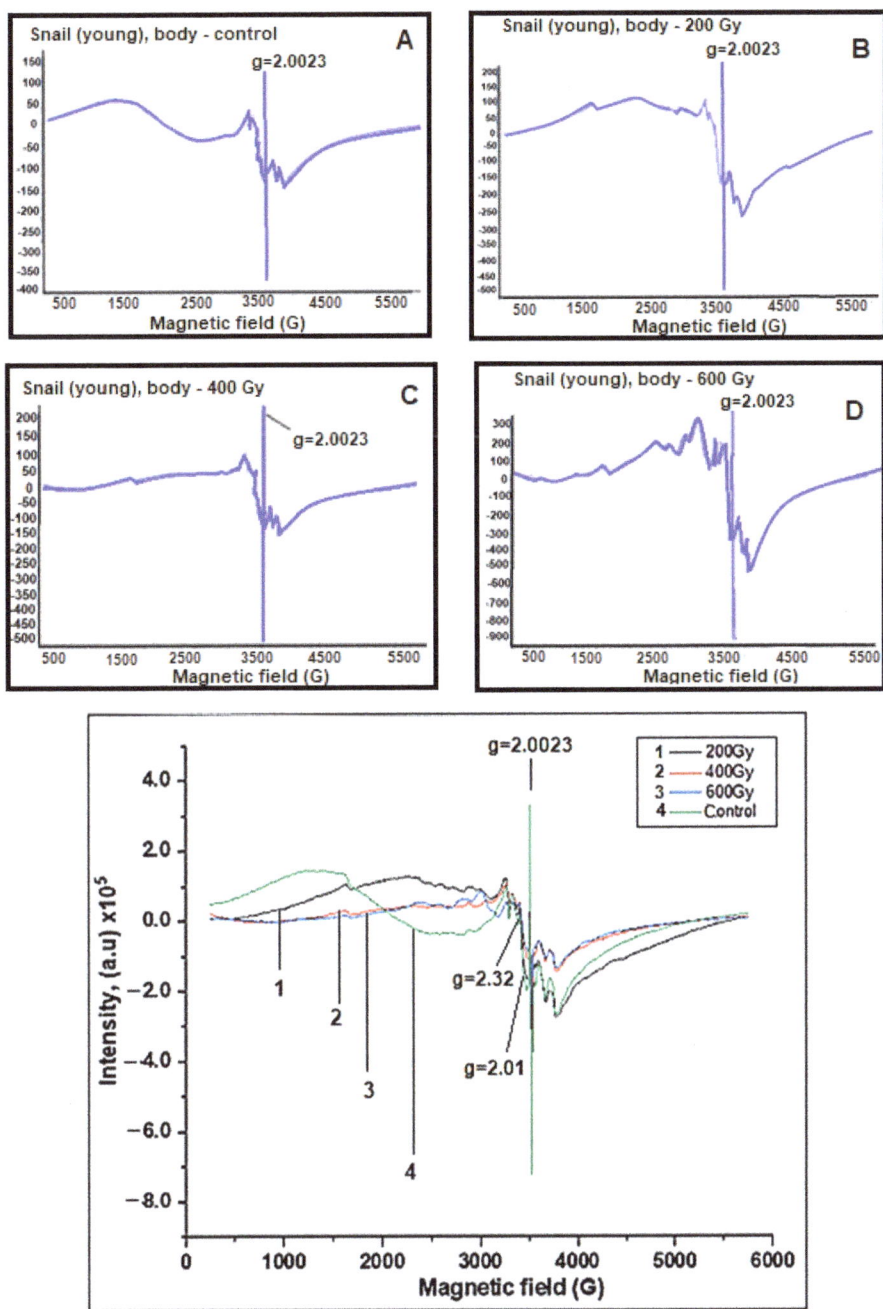

Figure 3. EPR spectra of body parts of the control and irradiated at different doses of young grape snails. (**A**) Control, (**B**) 200 Gy, (**C**) 400 Gy, (**D**) 600 Gy.

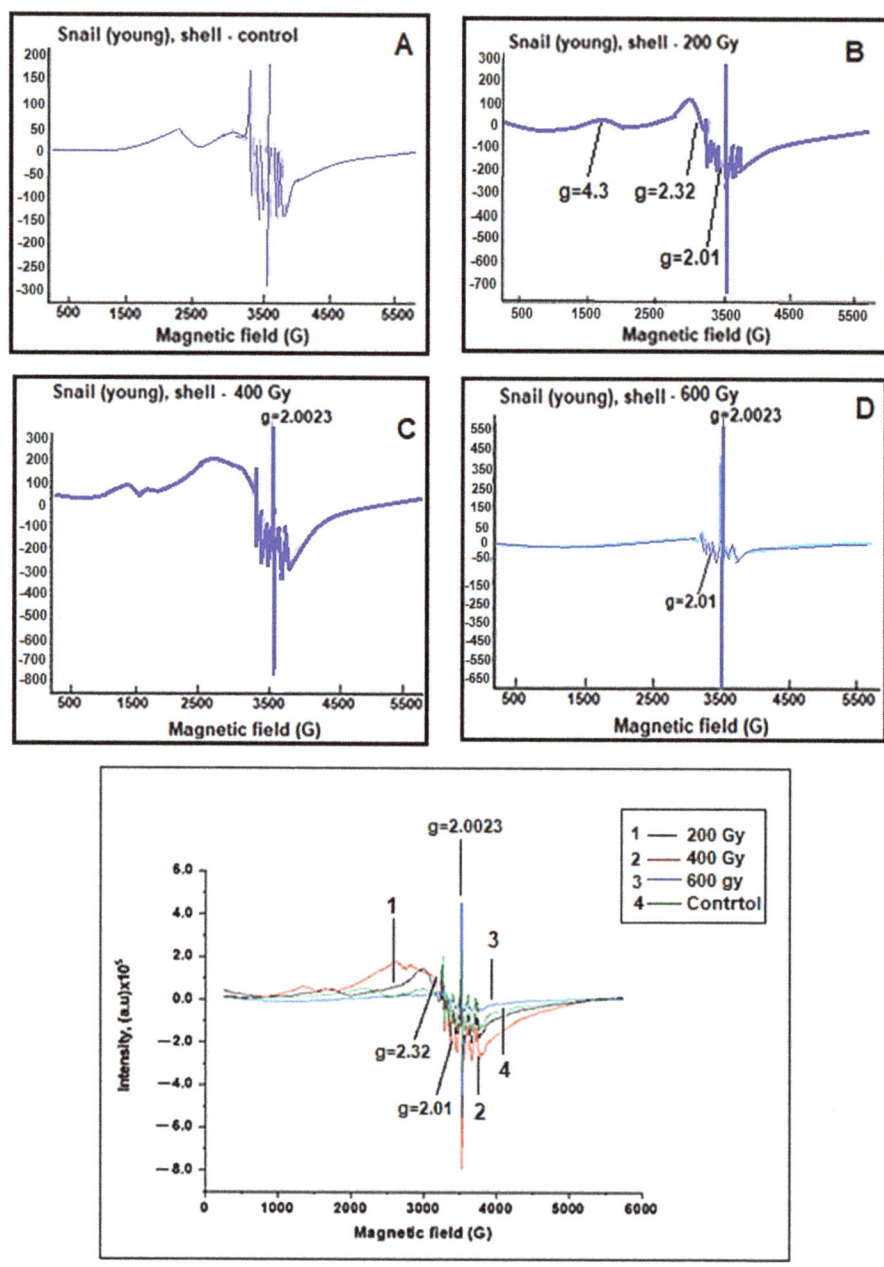

Figure 4. EPR spectra of shell parts of the control and irradiated with different doses of young grape snails. (**A**) Control, (**B**) 200 Gy, (**C**) 400 Gy, (**D**) 600 Gy.

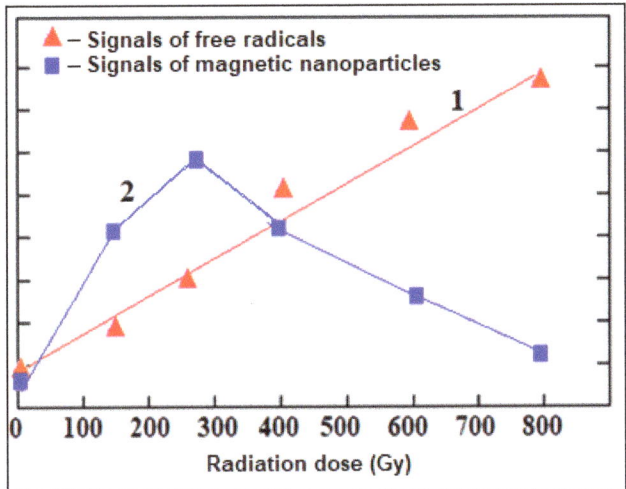

Figure 5. Dependence of the intensities of free radical signals (1) and broad EPR signals characterizing magnetic iron oxide nanoparticles (2) recorded in the body parts of grape snails on the radiation dose.

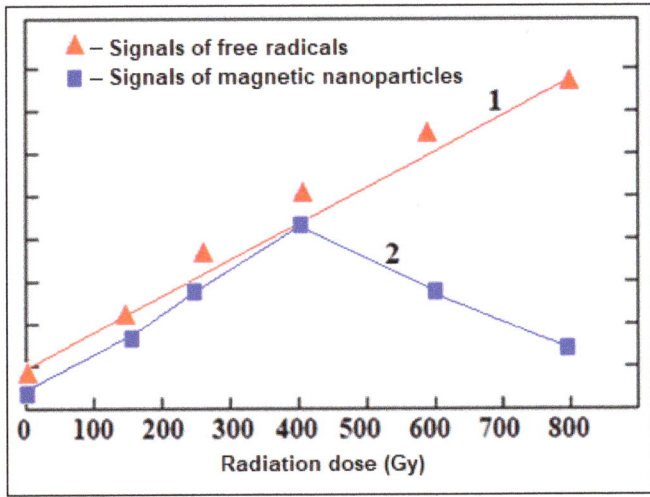

Figure 6. Dependence of the intensities of free radical signals (1) and broad EPR signals characterizing magnetic iron oxide nanoparticles (2) recorded in the shell parts of grape snails on the radiation dose.

As a continuation of the experiments, the radionuclide composition of the bodies and shells of the studied grape snails and their specific activities were also determined (Table 1).

It was found that more radionuclides (^{40}K, ^{232}Th, ^{226}Ra, ^{228}Ra, ^{137}Cs, ^{235}U, ^{238}U) are collected in the shell than in the body parts of grape snails. At the same time, higher specific activities of radionuclides were found in the shell. This can be explained by the fact that the shell is exposed to more stress factors than the body.

Table 1. Radionuclide compositions of shell and body parts of grape snails and their specific activities.

Radionuclides	Unit of Measure	Grape Snail Shell	Grape Snail Body
^{40}K	Bq/kg	26.1 ± 3.1	6.7 ± 1.5
^{232}Th	Bq/kg	6.2 ± 0.3	MDA = 0.32
^{226}Ra	Bq/kg	3.1 ± 0.5	MDA = 0.63
^{228}Ra	Bq/kg	4.3 ± 0.4	MDA = 0.55
^{137}Cs	Bq/kg	1.39 ± 0.12	0.91 ± 0.14
^{235}U,	Bq/kg	0.08 ± 0.03	MDA = 0.03
^{238}U	Bq/kg	1.65 ± 0.31	MDA = 0.65

Our research continued with the study of the mucin secreted by grape snails. Infrared spectra of mucin were recorded (see Figure 7).

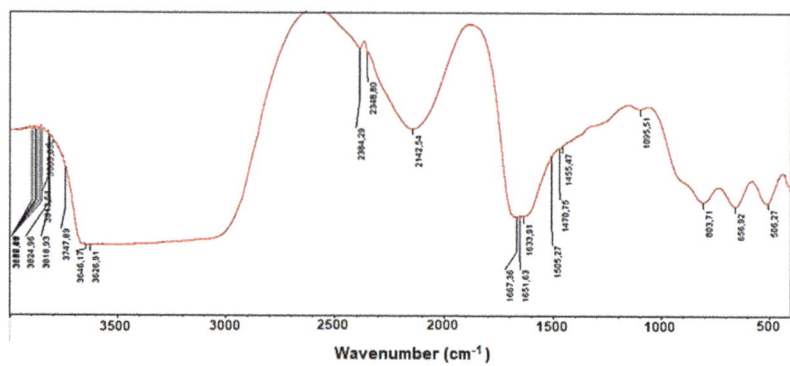

Figure 7. FTIR spectra of mucin secreted by grape snails.

In the spectra, the 3800–3300 wavenumbers belong to OH groups, 2142 to the absorption bands of C=C double bonds, 1667–1651 to the CO group, 803 (R)2 to the absorption band of out-of-plane deformation oscillations of C-H bonds connected to the double bond in the corner position of the C=CHR double bond, and 656, 508 to skeletal oscillations of C-C bonds.

3. Materials and Methods

In the conducted studies, young and old grape snails with shells collected from different areas of Absheron (Azerbaijan) were used as research objects. It should be noted that the age of snails is determined by the size of their shell parts. The shell color of grape snails is usually yellowish brown. Usually, they have wide stripes of dark brown color on their shells, but there are snails without them at all.

Snails are usually collected from nature in spring, summer, and early autumn. This is due to the hibernation of snails in cold seasons (at temperatures below 7 °C). Although the grape snail has a large shell and is slow, it can be a good digger. As soon as autumn comes, the snail digs a hole in the ground with its foot and then hibernates. If the ground cannot be dug due to the high density of the earth, the snail rolls over onto its back, scoops up more fallen leaves, and hibernates in this way. How long grape snails live is greatly influenced by their living conditions. In nature, this period is up to 8 years.

The grape snails that are the objects of research were collected from nature in spring and summer. After the snails were placed in special containers with 20 individuals in each, they were irradiated with different doses of ionizing gamma radiation (young snails:

200 Gy, 400 Gy, and 600 Gy; old snails: 50 Gy, 250 Gy, 450 Gy, 600 Gy, and 800 Gy) in a "RUHUND—20000" device with a CO 60 source (Figure 8). For 60 days after irradiation, the life activities and feeding of snails were monitored, and their death rates were determined. It should be noted that during this period, the food ration of the snails was completely the same. It was revealed that the life activities and nutrition of snails weaken with a high dose of gamma radiation. At the same time, the increase in the radiation dose led to an increase in their mortality rate. Thus, it was determined that two individuals irradiated with a dose of 200 Gy died, three individuals irradiated with a dose of 400 Gy died, and five individuals irradiated with a dose of 800 Gy died. In general, it was found that snails are resistant to the effects of radiation factors.

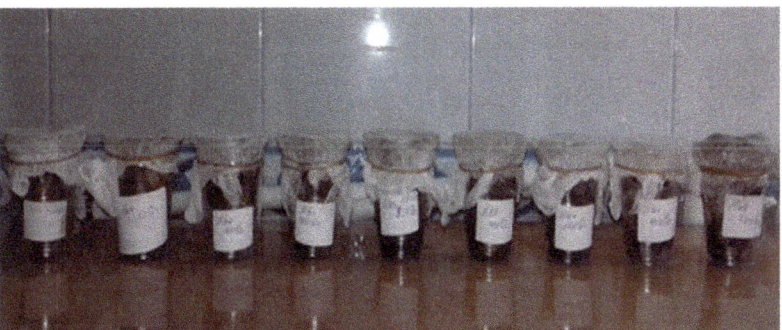

Figure 8. Control and irradiation with different doses of packaged grape snails.

After 60 days, the shells and body parts of the snails were separated and dried under natural conditions at room temperature (22–25 °C) (Figure 9) for 10–14 days. Spectra of the dried samples were recorded at room temperature on an Electron Paramagnetic Resonance Spectroscopy (EMX-BRUKER (Rheinstetten, Germany)).

Figure 9. Shell and body parts of grape snails dried at room temperature.

In addition, the mucin secreted by grape snails with shells was studied by the infrared (IR) spectroscopy method (Figure 7). The quality of the extracted mucin depends on many factors: temperature, season, and diet of the snail. The study of snail mucin is of great interest. The snails release mucin during times of stress or injury. As a complicated biological complex, mucin increases the regeneration properties of the epidermis, enriches the cells of the deep layers of the skin with water, fights inflammation, and acts as a highly effective natural antioxidant. Mucin is also used for cosmetic products.

The radionuclide composition of grape snails' bodies and shell parts and their specific activities were determined using CANBERRA gamma spectroscopy. This gamma spectroscopy is designed to measure the energies of X-ray or gamma radiation quanta emitted by radionuclides, as well as the activity (specific, volume) of gamma-emitting radionuclides in samples and objects. The results we obtained are shown in Table 1. For this purpose, when preparing the samples, the shell and body parts of the grape snails were separated from each other, dried at room temperature, and crushed by grinding. Then, the dried samples were placed in special Marinelli containers, and after some time (7–10 days), their radiospectrometric analysis was performed.

4. Conclusions

Using the EPR method, we studied new paramagnetic centers formed in living systems during the impact of various stress factors. The study of paramagnetic centers in various types of plant organisms has shown that the effect of stress factors causes the emergence of magnetic properties in them. In recent years, we have been conducting research on animal organisms to show the generality of the observed phenomenon. The presented work is dedicated to the study of new paramagnetic centers formed in old and young grape snails during the impact of ionizing gamma radiation, which is one of the stress factors.

The effect of ionizing gamma radiation on old and young grape snails was studied using the EPR method. The formation of paramagnetic centers was investigated in control grape snails and grape snails exposed to gamma radiation. It was found that gamma radiation causes the emergence of new magnetic properties in animal organisms, as well as plant systems.

EPR spectra of the body and shell parts of the control and exposure to various doses of gamma-irradiated young and old grape snails were recorded in a wide range of the magnetic field (500–5500 G). It was found that the intensity of the free radical signals ($g = 2.0023$) recorded in both the shell and body parts of grape snails increased linearly by elevating the gamma radiation dose. However, non-monotonic behavior of the intensities of the EPR signals ($g = 2.32$) characterizing the nanophase iron oxide particles was observed with the increase in the dose of gamma radiation in both the shell and the body parts. Therefore, the intensity of these signals was observed to increase gradually up to radiation doses of approximately 250–400 Gy, and gradually decrease during the subsequent increase of the radiation dose. In addition, it was determined that the amplitudes of the broad EPR signals characterizing iron oxide magnetic nanoparticles are more intense in the body parts of snails than in the shells. At the same time, a comparative study of old and young snails using the EPR spectroscopy method showed that the magnetic properties of old snails were higher than those of young snails. This can be explained by the fact that older snails are more exposed to various stress factors (temperature, UV radiation, drought, humidity, etc.).

The results obtained during the studies conducted with grape snails can be used in the assessment of the ecological status of the environment and many modern biomedical studies. The changes in the behavior of the EPR spectra of grape snails depending on the radiation dose show that these signals are informative in monitoring and evaluating the ecological state of the environment. Thus, depending on the dose of radiation, the regular change in the intensities of the free radical signals ($g = 2.0023$), as well as of the wide range of EPR signals characterizing iron oxide magnetic nanoparticles ($g = 2.4$) from the body and shell parts of snails, allows us to use snails as bioindicators.

The obtained results are also of practical importance in terms of applications in biomedicine, because in modern times, magnetic nanoparticles are widely used in medicine for both diagnostic and therapeutic purposes [33–35]. Additionally, we show in our research that magnetic properties are created in living systems during stress.

Thus, the study of paramagnetic centers in grape snails using the method of electron paramagnetic resonance spectroscopy showed that as a result of the impact of ionizing gamma radiation, which is one of the stress factors, nanophase magnetic iron oxide particles are formed in living systems.

It was determined that magnetite crystals in biological tissues are generated due to the phenomenon of biomineralization. The detection of iron oxide nanoparticles using

EPR signals can be used as a new source of biochemical and biophysical information in biomedical research.

Determining the radionuclide composition of the shell and body parts of grape snails separately, it was concluded that the radionuclides in the shell have a higher specific activity. This is explained by the fact that the shell of snails is more exposed to stress than the body.

Author Contributions: Conceptualization was performed by A.N. and R.K.; methodology was performed by M.B. and İ.M.; formal analysis was performed by M.A.; data curation was performed by A.N.; writing—original draft preparation was performed by A.N.; Visualization was performed by M.B., M.A. and İ.M.; writing, review, and editing were performed by A.E., A.N. and T.K.; investigation was performed by M.B., M.A. and İ.M.; supervision was performed by R.K., A.N., A.E., G.R. and D.S.; project administration was performed by R.K., A.N., G.R. and D.S. All authors have read and agreed to the published version of the manuscript.

Funding: This research received no external funding.

Institutional Review Board Statement: Not applicable.

Informed Consent Statement: Not applicable.

Data Availability Statement: The data used to support the findings of this study are included in the article.

Acknowledgments: This work was supported by the Faculty of Medical Sciences (JP 07/22), University of Kragujevac, Serbia. The authors are thankful for the moral support of the Institute of Radiation Problems, Baku, Azerbaijan. T.K. was supported by the Ministry of Education and Science of Ukraine (projects Nos. 0121U109543 and 0122U000874), National Research Foundation of Ukraine (project No. 2020.02/0100 "Development of new nanozymes as catalytic elements for enzymatic kits and chemo/biosensors"), and SAIA (Slovak Academic Information Agency) in the framework of the National Scholarship Programme of the Slovak Republic.

Conflicts of Interest: The authors declare no conflict of interest.

Sample Availability: Samples of the compounds are available from the authors.

References

1. Chelik, O.; Atak, C.; Suludere, Z. Response of soybean plants to gamma radiation: Biochemical analyses and expression patterns of trichome development. *Plant Omics* **2014**, *7*, 382–391.
2. Gudkov, S.V.; Grinberg, M.A.; Sukhov, V.; Vodeneev, V. Effect of ionizing radiation on physiological and molecular processes in plants. *J. Environ. Radioact.* **2019**, *202*, 8–24. [CrossRef] [PubMed]
3. Song, K.E.; Lee, S.H.; Jung, J.G.; Choi, J.E.; Jun, W.; Chung, J.W.; Hong, S.H.; Shim, S. Hormesis effects of gamma radiation on growth of quinoa (*Chenopodium quinoa*). *Int. J. Radiat. Biol.* **2021**, *97*, 906–915. [CrossRef]
4. Elgazzar, A.H.; Kazem, N. Biological effects of ionizing radiation. *Pathophysiol. Basis Nucl. Med.* **2006**, *23*, 540–548.
5. Zhang, H.; Zhao, Y.; Zhu, J.K. Thriving under stress: How plants balance growth and the stress response. *Dev. Cell* **2020**, *55*, 529–543. [CrossRef]
6. Eftekhari, A.; Arjmand, A.; Asheghvatan, A.; Švajdlenková, H.; Šauša, O.; Abiyev, H.; Ahmadian, E.; Smutok, O.; Khalilov, R.; Kavetskyy, T.; et al. The potential application of magnetic nanoparticles for liver fibrosis theranostics. *Front. Chem.* **2021**, *14*, 674786. [CrossRef] [PubMed]
7. Lukashev, E.P.; Oleinikov, I.P.; Knox, P.P.; Seifullina, N.K.; Gorokhov, V.V.; Rubin, A.B. The Effects of ultraviolet irradiation on hybrid films of photosynthetic reaction centers and quantum dots in various organic matrices. *Biophysics* **2017**, *62*, 722–727. [CrossRef]
8. Nasibova, A.N. UV-B radiation effects on electron-transport reactions in biomaterials. *Adv. Biol. Earth Sci.* **2022**, *7*, 13–18.
9. Nasibova, A.N.; Khalilov, R.I. Preliminary studies on generating metal nanoparticles in pomegranates (*Punica granatum*) under stress. *Int. J. Dev. Res.* **2016**, *6*, 7071–7078.
10. Khalilov, R.; Nasibova, A. The EPR parameter's investigation of plants under the influence of radiation factors. *Acta Bot. Caucasica* **2022**, *1*, 48–52.
11. Nasibova, A.N.; Trubitsin, B.V.; İsmailova, S.M.; Fridunbekov, İ.Y.; Qasımov, U.M.; Khalilov, R.I. Impact of stress factors on the generation of nanoparticles in the biological structures. *Rep. ANAS* **2015**, *71*, 35–40.
12. Khalilov, R.I.; Kavetskyy, T.S.; Serezhenkov, V.A.; Nasibova, A.N.; Akbarzadeh, A.; Davaran, S.; Moghaddam, M.P.; Saghfi, S.; Tkachev, N.A.; Milani, M.; et al. Detection of manganese-containing enzymes and magnetic nanoparticles in *Juniperus communis* and related biomaterials by ESR spectroscopy. *Adv. Biol. Earth Sci.* **2018**, *3*, 167–175.

13. Kavetskyy, T.S.; Khalilov, R.I.; Voloshanska, O.O.; Kropyvnytska, L.M.; Beyba, T.M.; Serezhenkov, V.A.; Nasibova, A.N.; Akbarzadeh, A.; Voloshanska, S.Y. Self-organized magnetic nanoparticles in plant systems: ESR detection and perspectives for biomedical applications. In *NATO Science for Peace and Security Series B: Physics and Biophysics*; NATO Advanced Study Institute (SPS. ASI 985310) on Advanced Technologies for Detection and Defence Against CBRN Agents; Petkov, P., Tsiulyanu, D., Popov, C., Kulisch, W., Eds.; Springer: Dordrecht, The Netherlands, 2018; pp. 487–492.
14. Nasibova, A.; Khalilov, R.; Eftekhari, A.; Abiyev, H.; Trubitsin, B. Identification of the EPR signals of fig leaves (*Ficus carica* L.). *Eurasian Chem. Commun.* **2021**, *3*, 193–199.
15. Nasibova, A.; Khalilov, R.; Abiyev, H.; Kavetskyy, T.; Trubitsin, B.; Keskin, C.; Ahmadian, E.; Eftekhari, A. Study of Endogenous paramagnetic centers in biological systems from different areas. *Concepts Magn. Reson. Part B* **2021**, *2021*, 6787360. [CrossRef]
16. Kavetskyy, T.S.; Soloviev, V.N.; Khalilov, R.I.; Serezhenkov, V.A.; Pan'kiv, L.I.; Pan'kiv, I.S.; Nasibova, A.N.; Stakhiv, V.I.; Ivasivka, A.S.; Starchevskyy, M.K.; et al. EPR study of self-organized magnetic nanoparticles in biomaterials. *Semicond. Phys. Quantum Electron. Optoelectron.* **2022**, *25*, 146–156. [CrossRef]
17. Khalilov, R.I.; Nasibova, A.N.; Kasumov, U.M.; Bayramov, M.A. Effect of radiation on wheat (Triticum L.) and corn (Zea mays L.): EPR studies. In Proceedings of the XXVIII International Conference "Mathematics. Computing. Education", Moscow, Russia 24–28 January 2022; p. 91.
18. Nasibova, A.N.; Khalilov, R.I.; Bayramov, M.A.; Bayramova, M.F.; Kazimli, L.T.; Qasimov, R.S. Study of some biophysical and biochemical parameters in stress—Exposed laboratory rats (Wistar albino). *J. Radiat. Res.* **2021**, *8*, 42–51.
19. Heybatova, N.; Nasibova, A. EPR studies of the effect of ionizing gamma radiation on Pelvic Grape Snails (*Helix pomatia* Linnaeus). In Proceedings of the XII International Scientific and Practical Conference, Topical Tendencies of Science and Practice, Edmonton, CA, USA, 7–10 December 2021; pp. 80–81.
20. Nasibova, A.; Kazimli, L.; Heybatova, N. Effect of metal nanoparticles on Pelvic grape snails (*Helix pomatia* L.). In Proceedings of the XVIII International Scientific and Practical Conference "Advancing in Research, Practice and Education", Florence, Italy, 8–11 March 2022; pp. 63–65.
21. Nasibova, A.N. The use of EPR signals of snails as bioindicative parameters in the study of environmental pollution. *Adv. Biol. Earth Sci.* **2019**, *4*, 196–205.
22. Andreev, N. Assessment of the status of wild populations of land snail (escargot) *Helix pomatia* L. in Moldova: The effect of exploitation. *Biodivers. Conserv.* **2006**, *15*, 2957–2970. [CrossRef]
23. Liu, Q.; Zhao, L.L.; Yang, S.; Zhang, J.E.; Zhao, N.Q.; Wu, H.; He, Z.; Yan, T.M.; Guo, J. Regeneration of excised shell by the invasive apple snail *Pomacea canaliculata*. *Mar. Freshw. Behav. Physiol.* **2017**, *50*, 17–29. [CrossRef]
24. Messina, L.; Bruno, F.; Licata, P.; Paola, D.D.; Franco, G.; Marino, Y.; Peritore, A.F.; Cuzzocrea, S.; Gugliandolo, E.; Crupi, R. Snail mucus filtrate reduces inflammation in canine progenitor epidermal keratinocytes (CPEK). *Animals* **2022**, *12*, 1848. [CrossRef]
25. Gugliandolo, E.; Macrì, F.; Fusco, R.; Siracusa, R.; D'Amico, R.; Cordaro, M.; Peritore, A.F.; Impellizzeri, D.; Genovese, T.; Cuzzocrea, S.; et al. The protective effect of snail secretion filtrate in an experimental model of excisional wounds in mice. *Vet. Sci.* **2021**, *8*, 167. [CrossRef] [PubMed]
26. Gugliandolo, E.; Cordaro, M.; Fusco, R.; Peritore, A.F.; Siracusa, R.; Genovese, T.; D'Amico, R.; Impellizzeri, D.; Di Paola, R.; Cuzzocrea, S.; et al. Protective effect of snail secretion filtrate against ethanol-induced gastric ulcer in mice. *Sci. Rep.* **2021**, *11*, 3638. [CrossRef] [PubMed]
27. Trapella, C.; Rizzo, R.; Gallo, S.; Alogna, A.; Bortolotti, D.; Casciano, F.; Zauli, G.; Secchiero, P.; Voltan, R. Helix Complex snail mucus exhibits pro-survival, proliferative and pro-migration effects on mammalian fibroblasts. *Sci. Rep.* **2018**, *8*, 17665. [CrossRef]
28. Nasibova, A.N.; Fridunbayov, İ.Y.; Khalilov, R.I. Interaction of magnetite nanoparticles with plants. *Eur. J. Biotechnol. Biosci.* **2017**, *5*, 14–16.
29. Khalilov, R.I.; Nasibova, A.N.; Gasimov, R.J. Magnetic nanoparticles in plants: EPR researchers. *News Baku Univ.* **2011**, *4*, 55–61.
30. Khalilov, R.I.; Nasibova, A.N.; Serezhenkov, V.A.; Ramazanov, M.A.; Kerimov, M.K.; Garibov, A.A.; Vanin, A.F. Accumulation of magnetic nanoparticles in plants grown on soils of apsheron peninsula. *Biophysics* **2011**, *56*, 316–322. [CrossRef]
31. Nasibova, A.N. The use of EPR signals of plants as bioindicative parameters in the study of environmental pollution. *Int. J. Pharm. Pharm. Sci.* **2015**, *7*, 172–175.
32. Nasibova, A.N. Formation of magnetic properties in biological systems under stress factors. *J. Radiat. Res.* **2020**, *7*, 5–10.
33. Flores-Rojas, G.G.; López-Saucedo, F.; Vera-Graziano, R.; Mendizabal, E.; Bucio, E. Magnetic nanoparticles for medical applications: Updated review. *Macromol* **2022**, *2*, 374–390. [CrossRef]
34. Tang, Y.D.; Zou, J.; Flesch, R.C.; Jin, T. Effect of injection strategy for nanofluid transport on thermal damage behavior inside biological tissue during magnetic hyperthermia. *Int. Commun. Heat Mass Transf.* **2022**, *133*, 105979. [CrossRef]
35. Koksharov, Y.A.; Gubin, S.P.; Taranov, I.V.; Khomutov, G.B.; Gulyaev, Y.V. Magnetic nanoparticles in medicine: Progress, problems, and advances. *J. Commun. Technol. Electron.* **2022**, *67*, 101–116. [CrossRef]

Disclaimer/Publisher's Note: The statements, opinions and data contained in all publications are solely those of the individual author(s) and contributor(s) and not of MDPI and/or the editor(s). MDPI and/or the editor(s) disclaim responsibility for any injury to people or property resulting from any ideas, methods, instructions or products referred to in the content.

Article

Heme Spin Distribution in the Substrate-Free and Inhibited Novel CYP116B5hd: A Multifrequency Hyperfine Sublevel Correlation (HYSCORE) Study

Antonino Famulari [1,2], Danilo Correddu [3], Giovanna Di Nardo [3], Gianfranco Gilardi [3], George Mitrikas [4], Mario Chiesa [2] and Inés García-Rubio [1,5,*]

1. Departamento de Física de la Materia Condensada, Universidad de Zaragoza, C/Pedro Cerbuna 12, 50009 Zaragoza, Spain; tonyfamulari@unizar.es
2. Department of Chemistry, University of Turin, Via Giuria 9, 10125 Torino, Italy; mario.chiesa@unito.it
3. Department of Life Sciences and Systems Biology, University of Turin, Via Accademia Albertina 13, 10123 Torino, Italy; giovanna.dinardo@unito.it (G.D.N.); gianfranco.gilardi@unito.it (G.G.)
4. Institute of Nanoscience and Nanotechnology, NCSR Demokritos, 15341 Athens, Greece; g.mitrikas@inn.demokritos.gr
5. Instituto de Nanociencia y Materiales de Aragón (INMA), CSIC-Universidad de Zaragoza, 50009 Zaragoza, Spain
* Correspondence: inesgr@unizar.es

Abstract: The cytochrome P450 family consists of ubiquitous monooxygenases with the potential to perform a wide variety of catalytic applications. Among the members of this family, CYP116B5hd shows a very prominent resistance to peracid damage, a property that makes it a promising tool for fine chemical synthesis using the peroxide shunt. In this meticulous study, we use hyperfine spectroscopy with a multifrequency approach (X- and Q-band) to characterize in detail the electronic structure of the heme iron of CYP116B5hd in the resting state, which provides structural details about its active site. The hyperfine dipole–dipole interaction between the electron and proton nuclear spins allows for the locating of two different protons from the coordinated water and a beta proton from the cysteine axial ligand of heme iron with respect to the magnetic axes centered on the iron. Additionally, since new anti-cancer therapies target the inhibition of P450s, here we use the CYP116B5hd system—imidazole as a model for studying cytochrome P450 inhibition by an azo compound. The effects of the inhibition of protein by imidazole in the active-site geometry and electron spin distribution are presented. The binding of imidazole to CYP116B5hd results in an imidazole–nitrogen axial coordination and a low-spin heme Fe^{III}. HYSCORE experiments were used to detect the hyperfine interactions. The combined interpretation of the gyromagnetic tensor and the hyperfine and quadrupole tensors of magnetic nuclei coupled to the iron electron spin allowed us to obtain a precise picture of the active-site geometry, including the orientation of the semi-occupied orbitals and magnetic axes, which coincide with the porphyrin N-Fe-N axes. The electronic structure of the iron does not seem to be affected by imidazole binding. Two different possible coordination geometries of the axial imidazole were observed. The angles between g_x (coinciding with one of the N-Fe-N axes) and the projection of the imidazole plane on the heme were determined to be $-60°$ and $-25°$ for each of the two possibilities via measurement of the hyperfine structure of the axially coordinated ^{14}N.

Keywords: EPR spectroscopy; CYP450; HYSCORE; peroxygenase; hyperfine interactions; low-spin hemeprotein; multifrequency EPR; quadrupole interaction; imidazole binding

1. Introduction

The cytochrome P450 family (CYP450s) consists of ubiquitous and versatile monooxygenases primarily responsible for catalyzing the hydroxylation of non-activated hydrocarbons using O_2 and NADPH, but they are also involved in numerous other reactions [1].

Existing in all living organisms such as human beings, animals, plants, bacteria, and fungi, CYP450s catalyze many important biological processes [2,3]. This great catalytic versatility, exploited in different ways to obtain high-value products such as steroids, fatty acids, and prostaglandins, as well as to eliminate xenobiotics and drug metabolites [4–6] has, over the years, attracted much interest in the scientific community. Additionally, CYP450s have also found use in the catalysis of specific reactions, such as epoxidation, desaturation, O, S, N-dealkylation, and sulfoxidation, as well as, if conveniently engineered, unnatural reactions [7–12]. Lately, CYP450s have also been reported to be key enzymes in cancer generation and treatment because they mediate the metabolic activation of numerous precarcinogens and participate in the inactivation and activation of anticancer drugs [13]. In this sense, azo compounds are part of several inhibitor-based drugs targeting CYP450s [14,15]. Finally, unwanted inhibition of CYP450s during drug metabolism and transport is one of the main causes of drug–drug interactions (DDIs) with consequent hospitalization and deaths related to drug use [16]. For these reasons, developing therapies allowing the possibility of controlling CYP450 inhibition is very attractive and could lead to substantial improvement in the treatment of certain medical conditions.

Among the members of the wide CYP450 family, CYP116B5 belongs to class VII CYP450 [17], which is defined as "self-sufficient" CYP450s [18] because they possess the P450 domain fused with the CYP450 reductase domain. Low activity, poor stability, and cofactor dependence are the main barriers to the industrial/biotechnological applicability of CYP450s and limit their use only to the production of high-value molecules such as fine chemicals and pharmaceuticals. Therefore, various efforts have been undertaken in recent decades to increase the application of CYP450s in biotechnology owing to enzyme engineering. For example, the heme domain (CYP116B5hd) obtained from the full protein CY116B5 showed outstanding resistance to H_2O_2 damage. This feature allows the enzyme to use the so-called peroxide shunt path to perform catalysis and, therefore, carry out different oxidative reactions on aromatic compounds and generate drug metabolites without economic or environmental drawbacks [19]. This pathway is a shortcut of the CYP450 catalytic cycle consisting in the use of hydrogen peroxide or peracids to generate, directly from the resting state, Compound 0, which precedes the formation of reactive Compound I, the true catalytic species of the whole catalytic cycle. Then, the latter can, through a stepwise or dynamically concerted radical rebound mechanism, immediately perform the hydroxylation reaction [20,21]. Therefore, without needing expensive electron donors, such as NADPH, used in the classical CYP450 catalysis, but simply by using hydrogen peroxide or peracids, products can be obtained with high catalytic performances [19,22].

The exceptional behavior of CYP116B5hd, which makes it different from a classical CYP450 monooxygenase, relies on the potential to exploit its peroxygenase-like reactivity [23]. With the conviction that this particular behavior had to be investigated by focusing on the properties and characteristics of the CYP116B5hd active site—the place where all the catalytic events take place—we undertook the study of the resting state of the enzyme via electron paramagnetic resonance (EPR) spectroscopy. Since CYP450s possess a characteristic active site with an Fe^{III}-heme center as the fulcrum, and given the paramagnetic nature of the iron, not only in the resting state of the enzyme but also in several intermediates generated during the CYP450 catalytical cycle, EPR spectroscopy lends itself as a very suitable technique to characterize and analyze this system [24–31]. In our previous work [32,33], we carried out a study on CYP116B5hd, where we analyzed the g-values according to the electronic model for low-spin Fe^{III} by Taylor [34,35] and compared our results with those obtained for other classical (monooxygenase-like) CYP450s, such as CYP102A1 (CYPBM3), and more exotic (peroxygenase-like) CYP450s, such as CYP152B1, CYP152K6, and CYP152L1. From this comparison, it was concluded that the electronic state of the enzyme was very much like the classic P450, determined by the active site in the close proximity of iron. Therefore, the peculiar behavior of this enzyme with respect to peroxide damage resistance would rather be associated with the supramolecular interactions between the protein scaffold and its active site. In the same study, the inhibitor imidazole was shown to bind the protein at the active site through direct coordination with the heme iron. The proof was the finding of an imidazole ^{14}N nucleus coupled with the electron spin of the iron using hyperfine sublevel correlation (HYSCORE) experiments. However, the complete analysis of nuclear spin frequencies obtained via HYSCORE spectroscopy or, in general, via hyperfine spectroscopy can still expose a wealth of information encoded in the hyperfine and nuclear quadrupole interactions. Such information is related to the geometry of the active site; for example, the location of a magnetic nucleus in virtue of its through-space magnetic interaction with the electron spin or related to the mapping of the electron spin density distribution [36–41]. This knowledge is relevant to enable structure–function relationships and to develop a molecular-level understanding of the factors governing the catalytic properties of heme-based enzymes.

This work, in particular, aims to clarify the coordination of the axial water molecule, since water molecules connected to the active site can shuttle proton channels and influence reactivity [42,43]. They definitely play an active role during the catalytic cycle of CYP450s, whether they electronically stabilize the Fe^{III} center in the resting state, favor the substrate approach to the active site, or the protonation of catalytic intermediates [44–46]. At the same time, there is the goal of conveying the geometric details of the active site and, especially, the orientation of the third iron orbitals where the unpaired electron resides and relating it with the coordination of the axial water molecule to find out the possible structural determinants of the electron distribution. Finally, the EPR spectroscopy characterization of CYP116B5hd is intended to be used as a model to study CYP450 inhibition by azo compounds and the effects of imidazole coordination on the structure of the heme.

In order to access this information, the complete hyperfine and nuclear quadrupole tensors have to be determined experimentally. We conducted this determination in a disordered sample (protein frozen solution) by taking advantage of the anisotropic g-tensor and systematically performing mostly HYSCORE experiments at different magnetic field values through the EPR spectrum, with special interest in the magnetic field positions that contain a principal axis of the g-tensor. Starting from there, we report on a detailed multifrequency (X-band and Q-band microwave frequencies) EPR spectroscopy investigation of the heme iron properties of CYP116B5hd, a self-sufficient monooxygenase acting as a peroxygenase, either in its resting state or interacting with imidazole, a basic model for azo compounds. This study was carried out by means of CW-EPR combined with hyperfine spectroscopy, providing a precise description of the electronic structure and environment of this peculiar cytochrome P450.

2. Results and Analysis

2.1. Effect of Nuclear Spin Labeling on the CW-EPR Spectrum

The CW-EPR spectrum of CYP116B5hd as a substrate-free protein either in an aqueous or deuterated frozen solution showed the typical powder pattern of an Fe^{III} low-spin ($S = \frac{1}{2}$) heme center (see Figure S1 in Supplementary Materials) with g-values (see Table 1) coinciding, within the error limits, with the ones previously reported [32].

Table 1. The g-values and crystal field parameters of the CYP116B5hd Fe^{III}-heme system in different experimental conditions.

Sample		g_x	g_y	g_z	V/ξ	Δ/ξ
CYP116B5hd in D$_2$O		2.443 ± 0.005	2.253 ± 0.002	1.923 ± 0.002	4.74 ± 0.06	5.44 ± 0.18
CYP116B5hd in H$_2$O [32]		2.440 ± 0.005	2.250 ± 0.002	1.920 ± 0.002	4.74 ± 0.06	5.44 ± 0.18
CYP116B5hd in H$_2$O + ^{15}N$_2$-imidazole	(2)	2.466 ± 0.005	2.258 ± 0.002	1.902 ± 0.002	3.47 ± 0.03	5.11 ± 0.17
	(1)	2.589 ± 0.005	2.258 ± 0.002	1.857 ± 0.002	4.40 ± 0.05	5.13 ± 0.17
CYP116B5hd in H$_2$O + Imidazole [32]	(2)	2.468 ± 0.005	2.258 ± 0.002	1.902 ± 0.002	3.50 ± 0.03	5.13 ± 0.17
	(1)	2.585 ± 0.005	2.258 ± 0.002	1.860 ± 0.002	4.39 ± 0.05	5.14 ± 0.17

In the presence of an excess of imidazole or $^{15}N_2$-imidazole, the CW-EPR spectrum is characterized by two new low-spin populations quantified to be approximately in the same amount, with neither one retaining the substrate-free protein g-values, as was found before in the protein–imidazole complex [32].

Labeling the nuclei of the axial distal ligand, H_2O or imidazole, with isotopes ^2H and ^{15}N, did not have any effect on the CW-EPR spectra, and consequently, on the crystal field parameters Δ and V. Therefore, one can conclude that the hyperfine couplings of the iron with these nuclei are smaller than other line-broadening mechanisms. To access the information about the hyperfine interactions occurring between the iron spin and the magnetic nuclei present in its environment, we previously tested the suitability of HYSCORE experiments, which showed couplings with several nuclei in the active site of the protein [32,33]. To fully characterize the anisotropy of the hyperfine and nuclear quadrupole interactions, HYSCORE experiments were recorded at different values of the magnetic field spanning the whole EPR spectrum of the protein, both with and without imidazole. To minimize the effect of blind spots, several τ values were used for the most significative spectra.

2.2. Hyperfine Interactions with Hydrogen Nuclei

The information related to couplings with hydrogen nuclei is observable only in the (+,+) quadrant, suggesting that these nuclei are weakly coupled to the iron electron spin. Among the closest protons to the iron are those belonging to the axial ligands, i.e., the two protons of the distal water molecule and the β (closest) or α protons of the proximal cysteine ligand (a.a. 381), and then the meso and pyrrole protons in the porphyrin ring.

In Figure 1, the (+,+) quadrants of the HYSCORE spectra best depicting the proton signals at the magnetic field position corresponding to the principal magnetic axes are shown.

All the proton signals emerge in the spectra as small ridges symmetrically placed with respect to the diagonal of the quadrant and located at or slightly above the antidiagonal crossing at the proton Larmor frequency for every magnetic field (dashed lines in the figure), consistent with proton signals in the weak-coupling regime ($A < 2\nu_H$). The lower range of frequencies in the 2D spectrum shows signals attributed to the ^{14}N nuclei, which are analyzed in the next section. The upper row of the figure (Figure 1a) collects the spectra from the resting state of the enzyme in an aqueous solution at the magnetic field positions corresponding to the g_z (left spectrum), g_y (center), and g_x (right).

For the g_z spectrum, three short ridges with slightly different directions are detected; they have been assigned to three different protons in the heme environment and labeled H_1, appearing at (15.9, 9.3) MHz, H_2 at (15.5, 9.5) MHz, and H_3 at (13.5, 10.9) MHz. It has been reported for CYP450s and other heme enzymes that the principal axis g_z is oriented approximately along the heme normal plane [47]. This means that the external magnetic field is perpendicular to the heme plane for this single-crystal position spectrum, which is an excellent opportunity to assign the proton signals using the electron–nuclear dipole–dipole interaction (Equation (1)). This approximation should be accurate enough when the interacting nucleus is located more than 0.25 nm away from the metal center and there is no significant spin delocalization out of the iron ion (see page 30 of [48]):

$$A_{\text{dip}} = \frac{\mu_0}{4\pi h} g_e \mu_B g_N \mu_N \frac{(3\cos^2\theta - 1)}{r^3} = T\left(3\cos^2\theta - 1\right) \quad (1)$$

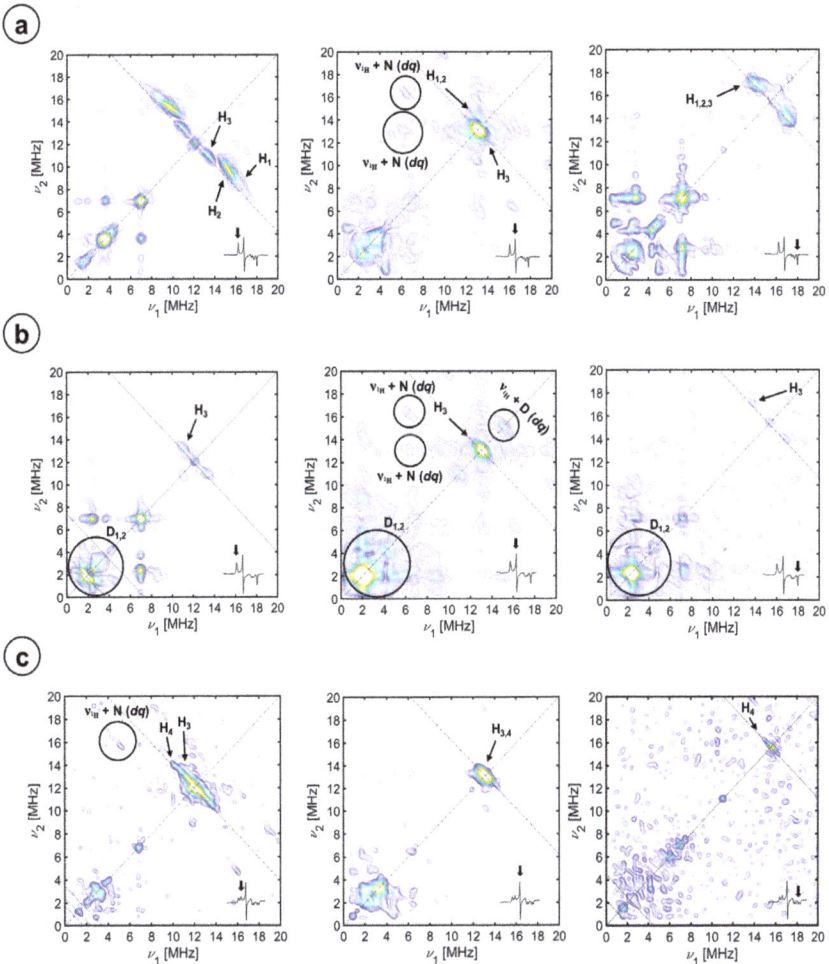

Figure 1. X-band, (+,+) quadrant, HYSCORE spectra of CYP116B5hd (300 μM) in KPi 50 mM pH 6.8, 30% glycerol substrate-free in H$_2$O (**a**), substrate-free in D$_2$O (**b**), and interacting with ^{15}N-imidazole (**c**). The spectra, as shown in the insets, were recorded at the (left column) g_z, (center column) g_y, and (right column) g_x magnetic field positions. The spectra were recorded at 10 K. τ values of (**a**) the sum of 208 ns and 250 ns spectra, 250 ns and 250 ns; (**b**) 400 ns, 250 ns, and 250 ns; and (**c**) the sum of 208 ns and 250 ns spectra, 250 ns and 168 ns.

The equation above is the expression of the hyperfine coupling between a nuclear and an electron spin with g_N and g_e, respectively. The distance between the two magnetic dipoles is represented by r, and θ is the angle between the vector $\vec{r}$ and the magnetic dipoles (oriented along the external magnetic field). For the HYSCORE spectra collected at g_z, the magnetic field is oriented along the heme normal plane, thus θ is the angle between the heme normal plane and $\vec{r}$ (see Figure 2).

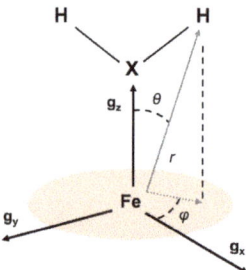

Figure 2. Sketch of Fe^{III}-heme center showing the geometric relationships between the g-tensor principal axes and a coupled proton nucleus belonging to an axial ligand and bound to a coordinated atom X.

Starting from the structural data of the closest protons to the iron, obtained from the X-ray diffraction study of the protein [49] (in Table 2), the position of the HYSCORE signals was calculated assuming exclusively a dipole–dipole hyperfine interaction with the electron density concentrated entirely on the position of the iron ion.

For cysteine beta protons (H$^\beta$), the information about r and θ was obtained directly from the structure; for the water protons, a range of compatible distances and angles was estimated from the position of the distal water oxygen. This simple calculation was sufficient to preliminary assign H$_1$ and H$_2$ signals to the two axial water protons and H$_3$ to one of the H$^\beta$ of the cysteine proximal ligand since the calculations were already giving results very close to the experiment. The water proton assignment of H$_1$ and H$_2$ was further confirmed by the lack of corresponding signals in the HYSCORE spectrum of the protein recorded under the same conditions but in a deuterated buffer, together with the appearance of symmetric peaks above the ^{2}H antidiagonal (see Figure 1b). The dipole–dipole couplings from other non-exchangeable protons in the active site, that is, the other H$^\beta$ proton of the proximal cysteine and from the porphyrin, were calculated to be too small to have a contribution to the signal outside the diagonal peak, especially considering that porphyrin protons have been reported to contribute to ENDOR spectra at this field position with couplings of 2 MHz [50–57]. With this argument, we confirm the assignment of H$_3$ signals to H$^\beta$ of the proximal cysteine ligand.

The HYSCORE spectra at the other single-crystal position, g_x, are depicted in the right column of Figure 1. Besides the matrix proton line at the diagonal, two short ridges are present along the ^{1}H antidiagonal at frequency coordinates (17.0, 14.1) MHz, which slightly move and change shape upon buffer deuteration. Concomitantly, a strong peak at the diagonal, close to the Larmor frequency of the deuterium appears. In this case, the magnetic field of the molecules contributing to the spectrum lies on the heme plane, in the direction of the principal x-axis of the g-tensor (which is in principle unknown).

The spectrum for the intermediate position g_y (central column in Figure 1) is contributed by molecules with different orientations of the magnetic field. The full range of orientations, containing the y-axis of the g-tensor, is depicted in Figure S1 (see Supplementary Materials). In addition to a strong peak in the diagonal, four peaks symmetrically placed with respect to the diagonal and the ^{1}H antidiagonal become evident together with weaker correlations forming a cross with the proton matrix peak at the center. These are combination peaks between the proton Larmor frequency (ν_H) and the double quantum frequencies of ^{14}N nuclei interacting with the electron spin (see below). Also, on the diagonal but slightly above the proton Larmor frequency, a small ridge is attributed to axial water protons based on their disappearance upon solvent deuteration (see Figure 1b, center).

The first step of the analysis procedure after the described assignment was to process the data according to Dikanov's procedure [58,59], as described in the Supplementary Materials. The correlation signals in the (ν^2_α, ν^2_β) plot are shown as straight lines, which indicates axial hyperfine interactions. Their linear fit allowed having a first estimate of the isotropic and axial anisotropic hyperfine contributions T and a_{iso}. These values were subsequently refined using HYSCORE simulations (see Figure 3) whereby the orientation of the anisotropic hyperfine was also determined.

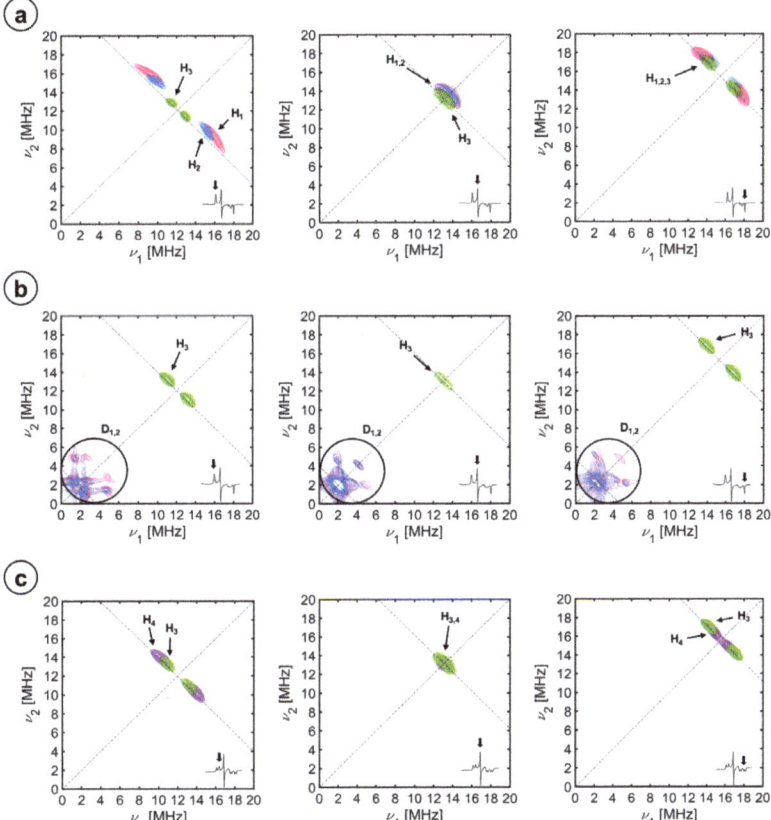

Figure 3. Simulation of the (+,+) quadrant, HYSCORE spectra of the substrate-free CYP116B5hd at the X-band. Exp. conditions: (300 µM) in KPi 50 mM pH 6.8, 30% glycerol in H_2O (**a**), substrate-free in D_2O (**b**), and interacting with ^{15}N-imidazole (**c**). The spectra, as shown from the insets, were recorded at the (left column) g_z, (center column) g_y, and (right column) g_x magnetic field positions. The individual simulations of the water protons, H_1 and H_2, are shown in pink and blue, those of the cysteine beta proton, H_3, in green, and the imidazole protons, H_4, in purple.

The angles α and β are the two Euler angles connecting the gyromagnetic axes with the hyperfine frame. Since the hyperfine tensor is axial, there is no need for the third Euler angle, γ. The parameters yielding the best simulations for the ^1H and ^2H signals of the complete set of spectra (shown in Figure 3) are collected in Table 2. Note that the angle β, which is the angle between the normal plane to the heme (directed as g_z in our case) and $\vec{r}$ and α, which is the angle between the g_x principal axis and the projection of $\vec{r}$ in the heme plane, corresponds to the angles θ and φ in Table 2 (see Figure 2).

Table 2. Spin Hamiltonian parameters of ^{1}H and ^{2}H proton nuclei coupled to the Fe^{III} electron spin, derived from simulations in Figure 3 on the spectra reported in Figure 1.

Species	Label	a_{iso} [MHz]	T [MHz]	EPR α, β [°]	EPR r (Fe-H) [Å]	Crystal str. r (Fe-H) [Å]	Crystal str. θ [°]	Crystal str. φ [°]
H$_2$O	H$_1$	−0.09 ± 0.06	5.60 ± 0.02	0 ± 5, 22 ± 5	2.42 [a]	2.9 [b]	23 [b]	5 [b]
H$_2$O	H$_2$	−1.095 ± 0.080	5.20 ± 0.02	0 ± 5, 16 ± 5	2.48 [a]	2.7 [b]	19 [b]	0 [b]
D$_2$O	D$_1$	−0.014 ± 0.01	0.860 ± 0.003	0 ± 5, 22 ± 5	2.42 [a]			
D$_2$O	D$_2$	−0.17 ± 0.03	0.800 ± 0.003	0 ± 5, 16 ± 5	2.48 [a]			
Cysteine	H$_3$	0.79 ± 0.22	2.60 ± 0.04	0 ± 5, 47 ± 5	3.12 [a]	3.078, 4.256 [49]	45, 66 [c]	4, 13 [c]
Imidazole (2)	H$_4$	1.76 ± 0.11	2.66 ± 0.08	−25 ± 5, 40 ± 5	3.10 [a]	3.142, 3.487 [60]	41, 38 [e]	N.A.
Imidazole (1)		1.76 ± 0.11	2.66 ± 0.08	−60 [d], 40 ± 5				

[a] Distances obtained through the point–dipole approximation (see Equation (1)). [b] Distances and angles obtained from the CYP116B5hd crystal structure (Res.: 2.60 Å) by tentatively adding the water protons using software. [c] Distances and angles obtained from the CYP116B5hd crystal structure, referring to the proximal cysteine, H$^\beta$. The farthest proton is not resolved from the matrix line in the HYSCORE spectrum. [d] α angle obtained from the spectral simulation of ^{14}N-Im g_x HYSCORE features since no proton signals are observed at the g_x of the Imidazole (1) species. [e] Distances and angles obtained from a CYP450$_{cam}$—imidazole complex (Res.: 1.50 Å, see [60]) and referring to both imidazole protons, H(2) and H(5).

From the value of T obtained from the simulations, the iron–proton distance was estimated (column 6 in Table 2) and compared with the distance and θ angle obtained from X-ray diffraction experiments (column 7). The agreement is remarkable. Note that the best simulations are obtained for α = 0 for water and cysteine protons, which means the projection of the respective r vectors coincides with the axis g_x.

The parameters for deuterium simulations were obtained by scaling the corresponding proton parameters by their nuclear gyromagnetic ratios. The simulations of the deuterium signals obtained using these hyperfine parameters together with a small nuclear quadrupole contribution of (0.1, −0.06, −0.04) MHz were satisfactory (see Figure 3). According to the crystal structure, the distance between the iron ion and the second H$^\beta$ of the proximal cysteine is 4.26 Å so the expected value for T would be around 1.03 MHz. The heme meso protons are located at 4.50 Å from the iron center and the heme pyrrolic protons have distances between 5.88 Å and 7.74 Å, corresponding to T values of 0.87 MHz and 0.39 MHz and 0.17 MHz, respectively. Such couplings were not detected in the HYSCORE spectra; they probably remain unresolved from the proton matrix line. To obtain very weak couplings, Mims ENDOR was performed (see Figure S4). They showed unresolved couplings within 1 MHz of the proton Larmor frequency, none of which was attributable to an exchangeable proton.

Upon inhibition by imidazole, the HYSCORE patterns change. To maintain the focus on the proton signals, the spectra shown in Figure 1c correspond to the protein with an excess of ^{15}N$_2$-imidazole, which has a much weaker axial nitrogen nuclear modulation. In the g_z spectrum, the proton signals assigned to the coordinated water molecule are absent in accordance with the replacement of the distal water ligand by the inhibitor. The proton signal that is actually detected consists of a ridge perpendicular to the diagonal. This signal overlaps with the correlation peaks detected in the substrate-free samples that were attributed to cysteine H$^\beta$; however, there is a substantial difference in the shape of the signal, which is more elongated along the ^{1}H antidiagonal. This difference is attributed to the additional contribution of 3- and 5-H imidazole protons. In the experimental spectrum at g_x (Figure 1, bottom right), the cross peaks assigned to the cysteine protons that are clearly visible in the deuterated sample could not be detected. On the other hand, a relatively intense peak on the diagonal newly emerges. At g_y, only a broad matrix peak on the diagonal is observed, which seems to be a bit more elongated along the diagonal than at the resting-state sample. The water proton signals vanished while the combination lines ($\nu_{1H} + {}^{14}N(dq)$) remained. Simulations of the 3- and 5-imidazole protons flanking the coordinating nitrogen were performed using the point dipole approximation with Fe-N-Im distances from other imidazole-coordinated heme proteins as the starting point [60] to search for the α angle to optimize the fit. This optimum value was −25°, which defines the orientation of the imidazole plane with respect to the axis g_x. The disappearance of the cysteine protons from the spectrum g_x is puzzling and could be due to a change in the direction g_x of the gyromagnetic tensor upon the addition of imidazole. We will investigate this instance carefully in the next section.

Note that the value of a_{iso} found for imidazole protons (1.76 MHz) is larger than those from water (0.09 MHz and 1.095 MHz) and cysteine protons (0.79 MHz).

2.3. Hyperfine Interactions with Nitrogen Nuclei

In Figure 4, selected X-band HYSCORE spectra of CYP116B5hd in the H$_2$O buffer showing the spectral regions with the contribution of ^{14}N heme signals are displayed.

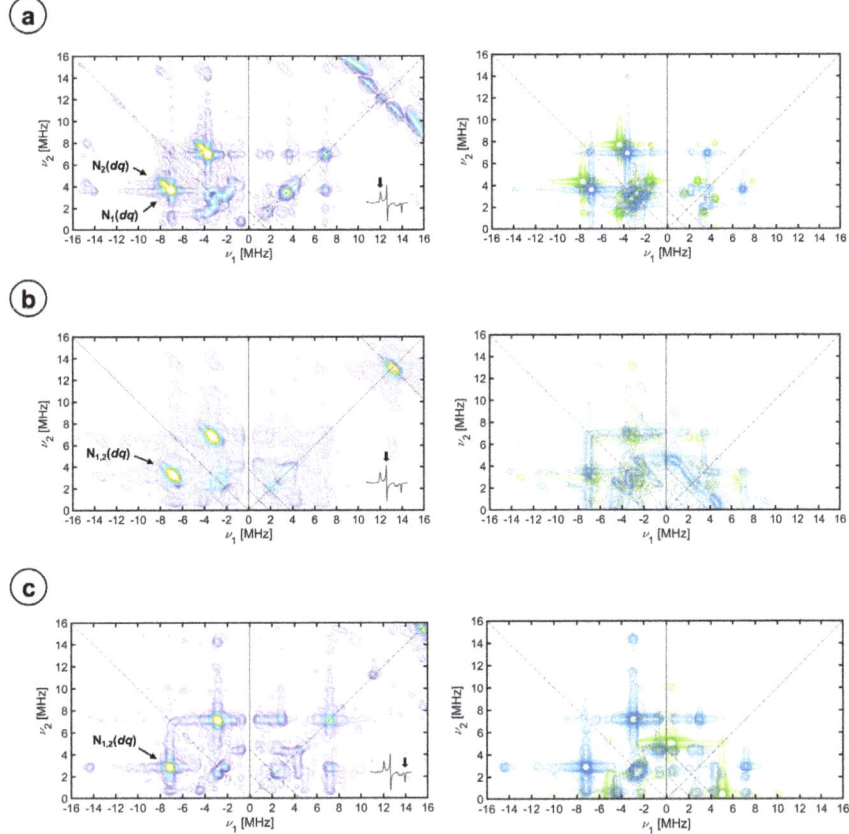

Figure 4. Experimental (**left**) and simulation (**right**) of X-band HYSCORE spectra of CYP116B5hd (300 μM) in KPi 50 mM pH 6.8, 30% glycerol substrate-free in H$_2$O. The spectra were recorded at the (**a**) g_z, (**b**) g_y, and (**c**) g_x magnetic field positions, at 10 K. τ values of (**a**) sum of 208 ns and 250 ns spectra, (**b**) 250 ns, and (**c**) 250 ns. The individual simulations of the heme ^{14}N pairs, N$_1$ and N$_2$, are shown in green and blue, respectively.

In general, the spectra are dominated by strong signals in the (−, +) quadrant. In all cases, the position of the more intense features lies on the two lines parallel to the diagonal that intercept the coordinate axes at 4 ν_N; that is, the difference between the nuclear frequencies in both spin manifolds is approximately four times the nuclear Larmor frequency of ^{14}N at the given magnetic field. This allows assigning these features as correlations of *double quantum* (*dq*) nuclear frequencies of two different ^{14}N nuclei (labeled as N$_1$ (*dq*) and N$_2$ (*dq*) in the figure) in the strong coupling regime (A > 2 ν_N). For the spectra corresponding to g_z, these peaks are found at (−7.00, 3.66) MH and (−7.61, 4.23) MHz, at (−6.77, 3.30) MHz for g_y, and at (−7.14, 2.89) MHz for g_x.

At frequencies below 5 MHz, there is a peak-dense region with correlation peaks involving *single quantum* (*sq*) nuclear frequencies that span both quadrants, (−,+) and (+,+).

The peaks observed at frequencies higher than 12 MHz in the (−,+) quadrant can be identified as combinations of nuclear frequencies of two nuclei. For the peak cluster around (−14, 4) MHz at the g_z spectrum, one of the coordinates is the sum of *dq* frequencies in one spin manifold (2·dq^{N1} and $dq^{N1} + dq^{N2}$) and the dq^{N1} or dq^{N2} nuclear frequency in the other spin manifold. Additional combination peaks are recognized above this cluster, for which the nuclear frequencies at both spin manifolds are the sum of *dq* frequencies. These combination peaks are also identified in the same region of spectra g_y and g_x, and they provide evidence of (at least) two equivalent nuclei of the one labeled N$_1$(*dq*).

Further experiments were performed at higher microwave frequencies in order to increase resolution, an equivalent set of Q-band HYSCORE spectra is shown in the Supplementary Materials (see Figure S5). For the Q-band, the hyperfine coupling is not in the strong regime but rather it is close to the exact cancellation condition A~2 ν_N for g_z and, therefore, the correlations appear in both quadrants. For g_x and g_y, the correlation peaks are found in the (+,+) quadrant since the interaction is, for the corresponding magnetic field value, in the weak coupling regimen (A < 2 ν_N). In the Q-band spectra, *dq* and *sq* correlations are observed, as labeled in the figure, but the s/n of the data did not allow recognition of combination peaks.

The simulations that are shown in Figure 4 (right) and Figure S5 of the Supplementary Materials were conducted by fitting all the spectra with a common set of parameters. The simulation routines were started by trying to minimize the number of varied parameters; therefore, reasonable assumptions were made such as taking the heme normal plane (g_z) as a principal axis for *A* and *Q* tensors of the nitrogen nuclei (see Equation (1)) [35–37,39], that is, the Euler angles for all tensors started as $\beta = 0°$ and $\gamma = 0°$. In general, the hyperfine coupling parameters were estimated from the *dq* signals and the nuclear quadrupole couplings from the *sq* peaks. Taking the axes of the *Q* tensor of the heme ^{14}N nuclei as the molecular axes, two groups of the ^{14}N nuclei were considered that were initially bound to differ by 90° in their orientation of the nuclear quadrupole axes with respect to the *g*-frame on the heme plane (Euler angle α). Complying with these assumptions, we were able to obtain the simulations shown in the figures without the need to release any of them. The simulation parameters are collected in Table 2. The hyperfine couplings of the two nitrogen nuclei differ only slightly, both hyperfine tensors are mostly isotropic, and the values are in the range of what has been reported before for other P450 enzymes and other low-spin heme centers [29,36–39,51,61]. While the isotropic contributions of the two sets of heme nitrogen nuclei are almost identical (N$_1$, a_{iso} = −4.93 MHz and N$_2$, a_{iso} = −5.10 MHz, see Section 3 for assignment of the sign), the anisotropic contributions are very small, a bit larger for the set N$_2$ (N$_1$, *T* = [0.035 0.135 −0.17] MHz and N$_2$, *T* = [0.30 0.40 −0.70] MHz). The traceless nuclear quadrupole tensors have principal values similar to those that have been reported [29,36,39] and the orientation in the heme plane coincides approximately with the N-Fe-N directions.

Upon imidazole addition, as discussed above, the formation of two imidazole-bound species takes place. For the EPR spectrum of this mixture of species, only the positions g_z and g_x of the species labeled Imidazole (1), the one with the most anisotropic EPR spectrum, is in a single crystal-like position. The positions g_z and g_x of Imidazole (2), the least anisotropic species have a more intense echo signal, and the HYSCORE spectra, with better s/n, contain the contribution of a single orientation of this species plus a set of orientations of the most anisotropic species. The spectra are depicted in Figure 5. They clearly differ from the ones recorded at the equivalent field positions of the enzyme resting state in an aqueous buffer. We have previously identified ^{14}N signals from the imidazole [32,33], but to make a complete assignment and analysis we prepared a sample by the addition of isotopically labeled ^{15}N-imidazole to the resting state of the protein. The corresponding HYSCORE spectra are shown for both species in Figure S6 of the Supplementary Materials and are all very similar to the ones obtained for the aqueous resting state at equivalent magnetic field positions and are well reproduced with its parameters (collected in Table 2). Additional weaker ridges present in the spectra are attributed to the heme nitrogen atoms of the other (most anisotropic) imidazole-bound species. These results evidence the lack of a substantial perturbation of the iron electron density upon imidazole binding and rule out a change in the orientation of g_x, which still coincides with one of the porphyrin N-Fe-N axes ($\alpha = 0$). Also, a detailed comparison of the aqueous and ^{15}N-imidazole bound spectra allowed the identification of weak ^{15}N correlation peaks along the diagonal parallel crossing at $2 \cdot \nu_{15N}$ for some of the spectra.

After studying the ^{15}N-Im and ^{14}N-heme signals, the signals from ^{14}N-imidazole were analyzed using the spectra of protein samples prepared with an excess of naturally abundant imidazole, evidencing drastic changes in the patterns of the single quantum transitions, at frequencies below 5 MHz. The dq correlation peaks of a new strongly coupled nitrogen nucleus, labeled N$_3$, at (-5.58, 2.40) MHz and (-5.96, 2.23) MHz frequencies at g_y and g_x magnetic field components of both Im-coordinated species are also visible in the spectra. The coupling tensors that best fit the imidazole ^{14}N signals are collected in Table 3 and the corresponding simulations are shown in Figure 5 on the right.

Table 3. Spin Hamiltonian parameters of ^{14}N and ^{15}N nuclei coupled to the Fe^{III} electron spin derived from simulations in Figures 4 and 5 of the experimental spectra reported in Figures S5 and S6.

Species	Label	A_x [MHz]	A_y [MHz]	A_z [MHz]	α, β, γ [°]	Q_x [MHz]	Q_y [MHz]	Q_z [MHz]	α', β', γ' [°]
Heme	N$_1$	-4.90 ± 0.1	-4.80 ± 0.1	-5.10 ± 0.1	$90 \pm 5, 0 \pm 5, 0 \pm 5$	0.90 ± 0.10	-0.60 ± 0.10	-0.30 ± 0.10	$90 \pm 5, 0 \pm 5, 0 \pm 5$
	N$_2$	-4.80 ± 0.1	-4.70 ± 0.1	-5.80 ± 0.1	$0 \pm 5, 0 \pm 5, 0 \pm 5$	1.00 ± 0.10	-0.60 ± 0.10	-0.40 ± 0.10	$0 \pm 5, 0 \pm 5, 0 \pm 5$
Imidazole (2)	N$_3$	$-3.57\ 0.1$	-3.20 ± 0.1	-2.54 ± 0.1	$65 \pm 5, 0 \pm 5, 0 \pm 5$	0.30 ± 0.10	0.80 ± 0.10	-1.10 ± 0.10	$65 \pm 5, 0 \pm 5, 0 \pm 5$
Imidazole (1)	N$_3$	$-3.57\ 0.1$	-3.20 ± 0.1	-2.54 ± 0.1	$30 \pm 5, 0 \pm 5, 0 \pm 5$	0.30 ± 0.10	0.80 ± 0.10	-1.10 ± 0.10	$30 \pm 5, 0 \pm 5, 0 \pm 5$
Imidazole-^{15}N$_2$	N$_4$	5.00 ± 0.1	4.48 ± 0.1	3.56 ± 0.1	$65 \pm 5, 0 \pm 5, 0 \pm 5$	N/A	N/A	N/A	N/A

For both species, the ^{14}N-Im hyperfine interaction was also found to be mostly isotropic (N$_3$, $a_{iso} = -3.10$ MHz) but somewhat weaker than the ones with the heme nitrogen nuclei. The small anisotropic contribution (N$_3$, $T = [-0.47\ -0.10\ 0.56]$ MHz) is small, in the same range as that observed for the porphyrin nitrogen nuclei; but, in any case, the total hyperfine contribution is less than the one reported for coordinated nitrogen atoms of imidazole in bis-histidine or histidine-methionine coordinated heme systems [36,38,62]. The nuclear quadrupole values are, however, very similar to the ones reported for imidazole in these systems [36]. The quadrupole parameters of ^{14}N-Im in many imidazole complexes have shown that the principal direction of a larger absolute value is the metal-^{14}N bond (for us, z) and the smallest one corresponds to the normal imidazole plane [63]. Therefore, taking the Q-tensor to indicate the orientation of the imidazole molecule, we determine that the orientation of the direction perpendicular to the imidazole plane with respect to the g-frame, which does not seem to be affected by imidazole coordination, is about 65° for Imidazole (2) and 30° for Imidazole (1). This explains why the g_z spectra are very similar for both species, whereas the g_x spectra differ substantially.

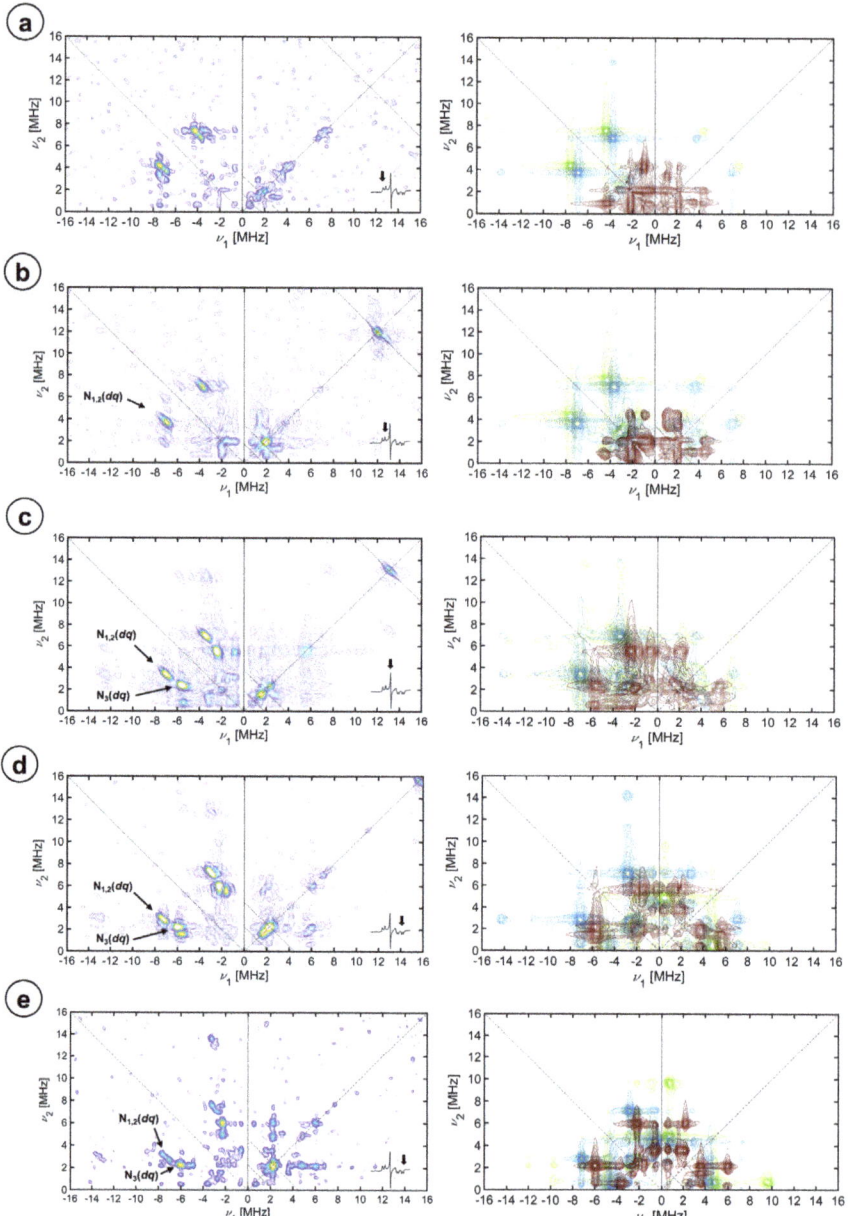

Figure 5. Experimental (**left**) and simulation (**right**) of X-band HYSCORE spectra of CYP116B5hd (300 μM) interacting with an excess of imidazole (1:10) in KPi 50 mM pH 6.8. The spectra were recorded at the magnetic field positions of (**a**) g_z (1), (**b**) g_z (2), (**c**) g_y, (**d**) g_x (2), and (**e**) g_x (1) at 10 K. τ values of (**a**) 208 ns, (**b**) 250 ns, and (**c**) 250 ns. The individual simulations of the heme ^{14}N pairs, N_1 and N_2, are shown in green and blue, respectively. The simulation of the imidazole ^{14}N is shown in dark red.

3. Discussion

In the CYP116B5hd active site, the Fe^{III} is coordinated by the four nitrogen atoms of the heme ring, forming a very stable complex; in addition to that, a water molecule and a cysteine residue act as axial ligands. The octahedral geometry and the strong ligands coordinating the Fe^{III} define a low-spin electron state ($S = \frac{1}{2}$), with the only unpaired electron of the ion dwelling in an orbital, which results from the mixing of the t_{2g} d orbitals caused by spin–orbit coupling [33–35,64,65]. Following this model, the g-values obtained from the CW-EPR spectra of the enzyme and of low-spin hemeproteins, in general, can be related to the energy of the t_{2g} orbitals of the iron considering a one-electron model (or rather a "hole"). The energy of the orbitals in a distorted octahedral environment can be parametrized by the axial crystal field parameter, Δ, and the rhombic crystal field parameter, V. Changes in the coordination environment of the Fe^{III}-heme center are reflected in a change in the relative energy of the orbitals and a change in the crystal field parameters in units of the spin–orbit coupling constant, ξ. As expected, isotope labeling of the axial water or imidazole does not affect the coordination environment reflected in Δ and V, which is driven by electronic interactions. However, the replacement of the axial water by imidazole causes a decrease in the value of Δ/ξ, which indicates a change in the strength of the ligand field caused by the change in the axial ligand. Both imidazole-coordinated species have, within the error limits, the same Δ/ξ parameter. The V/ξ values, accounting for the lack of axiality in the spin system since it measures the energy difference between the d_{zx} and d_{zy} orbitals, are different for the two species, which, as we demonstrate here, is due to the imidazole plane adopting a different orientation with respect to the porphyrin ring [32,33].

The CW-EPR analysis (see Section 2.1) indeed helped to characterize the electron density distribution within the t_{2g} orbitals and understand changes in the coordination geometry of the iron; however, the CW spectra do not allow studying the weak interactions between the iron unpaired electron spin and nuclei close by because the hyperfine structure is not resolved. Since further valuable information about the active site is contained in these interactions, the Fe^{III}-heme system in the resting state and inhibited with imidazole was meticulously studied with HYSCORE, leading to the complete characterization of the hyperfine and nuclear quadrupole tensors of protons and nitrogen nuclei in the close environment of the electron spin.

The Dikanov methodology allowed us to determine that the proton hyperfine coupling tensors were axial and to obtain an estimation of a_{iso} and T for the proton signals visible in the HYSCORE spectra that were later refined with simulations. From the value of the axial anisotropic hyperfine parameter T determined with this method and using the spin–nuclear dipole–dipole approximation (see Equation (2)), the distances between these nuclei and the unpaired electron of iron were calculated. These distances are highly consistent with the ones obtained from the crystal structure (see Table 1). Also, the Euler angle β was found to be very close to the angle θ measured between the heme normal plane and the vector $\vec{r}$ for every nucleus. This fact confirms that the principal axis associated with g_z is directed along the heme normal plane. Additionally, the angle φ, which is the angle between the projection of the Fe-H bond onto the heme plane and the direction of g_x, was obtained for every proton by simulating the set of experimental data (this angle is the Euler angle α). The combination of φ values obtained for all protons allowed us to locate the direction of the g_x axis itself, which coincides with the direction N (A pyrrole)-Fe-N (C pyrrole) of the heme plane. For the axial water molecule, we were able to resolve two different proton nuclei with very similar anisotropic couplings (~5 MHz), which is the main contribution to the hyperfine tensors for both protons but differs in the small isotropic hyperfine contribution a_{iso}. The values of the distance estimated using the point dipole approximation are ~2.4 Å, which are consistent with what has been observed before for other CYP450s with a difference of 0.2 Å [52,66]. From the α angles of the hyperfine tensors, found to be 0 for both water protons, one can conclude that the plane of the water molecule is perpendicular to the heme plane lying on the g_x axis. The distances and theta angles found for each of the two protons were very similar, which, taking into account that the two protons are in the same water molecule, could be interpreted as the coordination bond of Fe-O between the axial ligands and the iron by approximately bisecting the H-O-H angle (see Figure 6) as has been observed in previous studies [52,66].

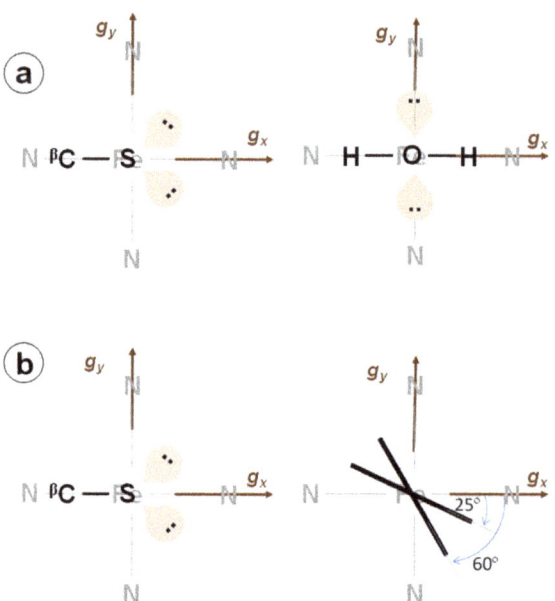

Figure 6. Sketch of the active site of CYP116B5hd. Views from the proximal (left) and distal sites (right). (**a**) CYP116B5hd in the resting state. (**b**) Imidazole inhibited CYP116B5.

The possibility of the two water protons belonging to two slightly different conformations of the axial water cannot be ruled out with the presented evidence; however, this result is not preferred since we think the extreme sensitivity of the CW-EPR spectrum to the conformation of the axial ligands would have shown two resolved species in the spectrum. As can be seen from Figure 1b, the presence of deuterated water in the active site is confirmed by the disappearance of the two external ridges (H_1 and H_2) from the proton signal pattern and the appearance of deuterium signals with hyperfine couplings that correspond perfectly to the reported values for the proton hyperfine tensors once the appropriate scaling is carried out according to their different nuclear Larmor frequencies. The nuclear quadrupole coupling observed is very small and consistent with what was observed before for water in a Fe-S cluster protein [67].

The signals from the cysteine β proton are consistent with those found for other CYP450s [30,40,66]. Again, the Fe-$H^{β\text{-Cys}}$ distance obtained from the point dipole approximation of the hyperfine coupling and the tensor orientation allowed understanding that, from the two beta protons, the observed couplings originate from the one directed toward the heme, which is closer to the iron. Of course, the signals assigned to this proton do not disappear from the spectra upon solvent deuteration since only the water protons are exchangeable protons (pKa = 15.7); those of the cysteine, being an alkane-like species (pKa = 50), are not. The angle α obtained for this proton is $0 \pm 5°$, which indicates that the projection of the Fe-beta proton direction approximately coincides with g_x (see Figure 6). Structural XRD studies of the protein show that this projection is 4° from the N (A pyrrole)-Fe-N (C pyrrole) direction. The cysteine proton signal that was observed in the g_x spectrum should remain after the binding of imidazole unless the g_x direction changes. We determined that this last hypothesis is not compatible with the observed nitrogen hyperfine structure; therefore, we conclude that for some unknown reason, we failed to detect these signals that lie below the s/n ratio of the spectra.

Parallel studies were carried out by means of the ENDOR technique, which is another useful method that is able to detect nuclei coupled to the unpaired electron; this method could have confirmed what was observed with HYSCORE. However, the results were unsatisfactory in detecting the near axial water molecule proton. The Davies ENDOR sequence yielded a very low s/n and hardly any signal was discernible. Regarding the distant heme peripheral protons (meso/pyrrolic) of the porphyrin ring, the Mims ENDOR showed only unresolved signals of non-exchangeable protons corresponding to couplings of less than 2 MHz.

The proton signals originating from the imidazole were never completely resolved from the ones of cysteine H^β in the HYSCORE spectra. The distances and θ angles obtained from the HYSCORE spectrum (note that there is no crystal structure of the Im-inhibited protein), correspond to what was seen before for axially coordinated imidazole cytochromes [60]. The α angle of 65° obtained for the species Imidazole (2) is interpreted as the orientation of the imidazole plane relative to the g_x direction (coinciding with the N_A-Fe-N_C direction). The hyperfine couplings between the iron electron spin and the nitrogen nuclei also yield important structural information and further details about the electron spin localization in the active site of CYP116B5hd. In the resting state of the enzyme, the only nuclei strongly coupled to the metal are the ^{14}N of the porphyrin covalently bound to the iron ion. The experimental spectra show two sets of non-equivalent nuclei that we call N_1 and N_2. This non-equivalency between the heme ^{14}N has previously been observed in CYP450$_{cam}$ or low-spin heme complexes [39], although they have been reported as equivalent in other ferric low-spin systems [29,62]. The various combination cross peaks emerging at the sum of dq nuclear frequencies that are detected in the different spectra led us to the conclusion that the signals of N_1 originate from at least two magnetically equivalent nuclei.

The whole set of HYSCORE spectra at X- and Q-band frequencies could be reasonably simulated with the hypotheses that (1) the direction perpendicular to the porphyrin plane, as a symmetry axis, is approximately a principal axis of the hyperfine and nuclear quadrupole tensors for ^{14}N coordinated nuclei and (2) the two sets of nitrogen nuclei correspond to the two sets of diametrically opposed heme nitrogen atoms and therefore their Euler α angles differ by 90°. Table 3 collects the parameters used for the simulations of the experimental data sets of nuclear frequency patterns (both shown in Figure 4). For both sets of nitrogen nuclei, N_1 and N_2, the hyperfine tensor is predominantly isotropic with principal values that are in the range of those found for pyrrole nitrogen nuclei of other ferric heme proteins [31,38–41,53,63]. The observed lack of magnetic equivalence of the four heme ^{14}N ligands is basically due to a small but measurable difference in the hyperfine coupling constant along the heme normal plane, which is a bit larger for N_2. This effect must be due to the breaking of the heme symmetry by distortions affecting the heme. Indeed, in the crystal structure at room temperature, the heme is reported to suffer from a small ruffling distortion: (in absolute values) 0.30 for CYP116B5hd, 0.38 for P450$_{cam}$, and 0.09 for human aromatase, calculated using the PyDISH online tool at https://pydish.bio.info.hiroshima-cu.ac.jp/, accessed on 18 January 2024 [68].

Nitrogen combination cross-peaks similar to the ones observed here have been observed in bis-histidine low-spin heme systems which, together with an additional combination peak between a coupled proton and a heme nitrogen, allowed us to assign in this system a negative sign to the hyperfine coupling of the ^{14}N coordinated nuclei [36] (heme and imidazole in that case). Based on the similarities of the two systems, we tentatively assign a negative sign to the hyperfine coupling constants of heme nitrogen atoms in CYP116B5hd.

The nuclear quadrupole tensors obtained from our data are close to the ones reported for similar proteins [36,39]. The fact that the Q-tensor principal values are very similar, with a β Euler angle of 0° and an α angle close to 0° for one of the nitrogen sets (N_2) and

90° for the other (N_1) set, supports the premise that both sets are chemically equivalent but their spatial arrangement in the heme plane are rotated by 90°. Taking into account that the principal axes of the nuclear quadrupole tensors follow the molecular symmetry directions at the nucleus and that the larger principal value for pyrrole heme nitrogen nuclei has been reported to be perpendicular to the Fe-N direction in the heme plane, we can assign the set N_1 to the nitrogen atoms, whose N-Fe-N direction is aligned with the g_x direction ($\alpha \sim 90°$), and the set N_2 to the other set of heme nitrogen atoms, whose N-Fe-N would be aligned with g_y. Therefore, from both the heme nitrogen and the cysteine H^β nuclear frequency data, we can conclude that the direction of the g_x axis (and therefore the g_y axis too) coincides with the experimental uncertainty with the N-Fe-N directions of the heme. According to the counter-rotation theory [35], this means that the t_{2g} orbitals are also oriented along the heme axes. One of the structural asymmetries of the axial ligands that are expected to split the energy of the orbitals d_{zx} and d_{zy} is the orientation of the second lone pair of sulfur in the cysteine. In the crystal structure, the projection of this lone pair onto the heme plane also approximately coincides with one of the heme axes (4°). Moreover, according to the water proton data presented here ($\alpha \sim 0°$), the axial water arrangement also follows the heme directions. This makes sense since the axial water is free to rotate to attain the minimal energy configuration. In conclusion, from our experimental data, it seems that the porphyrin N-Fe-N bonds and the cysteine command the orientation of the Fe^{III} electron density and the axial water accommodates it.

The addition of the inhibitor strongly changes the appearance of the nitrogen HYSCORE spectrum in all the magnetic field positions. This is due to the presence of an extra nitrogen nucleus, which directly and strongly coordinates with the metal center. The inhibition of the protein with imidazole isotopically labeled with ^{15}N allows for studying the effect of imidazole on the hyperfine structure of heme nitrogen nuclei and cysteine protons since the echo modulation amplitude due to the axial ^{15}N is very small and hardly visible in the spectra. The data show that imidazole coordination does not disturb the hyperfine couplings for any of the imidazole-coordinated species. The large echo modulation amplitude due to the imidazole ^{14}N nuclei allows for identifying its cross-peaks in all HYSCORE spectra and leads to the determination of the full hyperfine and nuclear quadrupole coupling tensors of this new ^{14}N nucleus (Table 3). The hyperfine values are smaller (in absolute value) than those found for the iron-binding bis-His or His-Met heme centers, [36] but agree with those reported for the aromatase–anastrozole complex [29], which has cysteine–azo coordination. Moreover, hyperfine values of about 2 MHz have been reported for the axial amine's nitrogen in imidazole–heme–mercaptoethanol and in pyridine–heme–mercaptoethanol complexes [69,70].

Additionally, according to a ^{14}N and ^{2}H quadrupole double resonance study of imidazole and substituted derivatives, the largest value of the nuclear quadrupole tensor in absolute value, $|Q_z| = 1.1$ MHz, is an indication that the coordination occurs via the N(3) of the imidazole ring [71]. The difference of this value with respect to the one reported, $|Q_z| = 1.61$ MHz, is explained by the fact that the addition of a Lewis acid (Fe^{III}) to an amine (imidazole) leads to a reduction in the electric field gradient at the nitrogen nucleus [72]. This value is consistent with nuclear quadrupolar tensors measured for the N(His) of the myoglobin–mercaptoethanol complex and metal-coordinated nitrogen nuclei in different ligands, including imidazole, studied via nuclear quadrupole resonance [63,72]. The important feature to be emphasized from these studies is that nuclear quadrupole values of the order of those measured here for imidazole ^{14}N are associated with the lone pair nitrogen donor orbital defining the principal axis of the nuclear quadrupolar coupling tensor (Q_z). This orbital is axially directed toward the empty iron d_z^2 orbital forming an s bond along the heme normal plane. Simulation of the HYSCORE spectra demonstrates that Q_z is nearly aligned along g_z ($0° < \beta < 10°$), indicating that the axial nitrogen ligand is directed close to the heme normal plane. The Euler angle alpha of the ^{14}N-imidazole nuclear quadrupole tensor indicates the projection of the direction perpendicular to the imidazole plane onto the heme plane. The HYSCORE patterns show that this orientation is 65° for the least anisotropic imidazole-coordinated species (Imidazole (2)), which is in full agreement with what was found in analyzing the imidazole proton signals, where the alpha angle indicates the orientation of the projection of the imidazole plane onto the heme plane. On the other hand, the orientation perpendicular to the imidazole plane was found to be closer to 30° for the second species (Imidazole (1)), which substantiates the differences between both species as hypothesized based on the crystal field parameter V/ε. No cysteine proton signals were found but, interestingly, the hyperfine parameters of the ^{14}N-heme ligands remained the same for the resting state and Imidazole (2), which confirms that heme and cysteine control the electronic structure of the iron.

4. Materials and Methods

Cloning, expression, and purification of CYP116B5hd. The construct used in this work was obtained by cloning the initial part of the gene of CYP116B5 (coding for the first 442 amino acids, the heme domain) between NdeI and EcoRI restriction sites in a pET-30a(+) vector with the insertion of a N-terminal 6xHis-tag [19]. Expression and purification of the protein were carried out as previously described in detail in [19]. Briefly, protein expression was carried out in *E. coli* BL21 (DE3) cells at 22–24 °C for 24 h in LB medium supplemented with 0.5 mM of δ-aminolevulinic acid (δ-Ala) and 100 μM of IPTG. For the purification process, the cells were resuspended and sonicated in 50 mM of KPi buffer, pH 6.8, supplemented with 100 mM of KCl, 1 mg/mL of lysozyme, 1% of Triton X-100 and 1 mM of PMSF (phenylmethylsulphonyl fluoride), and 1 mM of benzamidine. After ultracentrifugation at 90,000× *g* for 45 min at 4 °C, the supernatant was loaded onto a 1 mL His-trap HP column (GE Healthcare, Chicago, IL, USA) and eluted using a linear gradient of imidazole ranging from 20 to 200 mM. The purest fractions were then concentrated, loaded into a Superdex 200 size exclusion chromatography column (GE Healthcare, Chicago, IL, USA), and eluted using 50 mM of KPi buffer, pH 6.8, containing 200 mM of KCl. The purified protein was then concentrated and stored in 50 mM of KPi buffer, pH 6.8, containing 10% glycerol after buffer exchange via ultrafiltration using Amicon Ultra 30,000 MWCO devices (Merck, Darmstadt, Germany). Deuterated protein samples were prepared by exchanging the aqueous buffer with 50 mM of KPi buffer, pH 6.8, containing 10% glycerol, prepared in D2O. Protein concentration was estimated from the spectrum of the P450−CO complex upon reduction with sodium dithionite and CO bubbling, using an extinction coefficient of 91,000 M^{-1} cm^{-1} [73].

Electron Paramagnetic Resonance. All the protein samples in 50 mM of KPi buffer, pH 6.8, with 10% glycerol were mixed with 30% glycerol as a glassing agent to an approximate final protein concentration of 200 μM. The samples with imidazole (natural abundance, hereinafter called simply imidazole) or $^{15}N_2$-imidazole were prepared adding an excess of this chemical to reach the ratio (1:10) with respect to the protein. X-band experiments were performed on a Bruker ELEXSYS E580 spectrometer (microwave frequency 9.68 GHz) equipped with a cylindrical dielectric cavity and a continuous helium gas flow cryostat from Oxford Inc. (Atlanta, GA, USA). Q-band Pulse EPR experiments were performed on a Bruker ELEXSYS E580 spectrometer (microwave frequency 34 GHz) equipped with a continuous helium gas flow cryostat from Oxford Inc. (Atlanta, GA, USA). The magnetic field was measured by means of a Bruker ER035M NMR gaussmeter (Bruker, Ettlingen, Germany).

Continuous-Wave EPR. The experiments were performed on the Bruker ELEXSYS E580 X-band spectrometer at a temperature of 40 K. A microwave power of 0.31 mW, a modulation amplitude of 0.6 mT, and a modulation frequency of 100 KHz were used.

Hyperfine Sublevel Correlation (HYSCORE) [74]. Pulse EPR experiments were performed at 10 K using the pulse sequence $\pi/2$-τ-$\pi/2$-t_1-π-t_2-$\pi/2$-τ-echo with microwave pulse lengths $t_{\pi/2}$ = 16 ns and t_π = 16 ns. The time intervals t_1 and t_2 were varied in steps of 16, 24, or 48 ns. To avoid overlooking correlation peaks due to blind spot effects, the experiments were performed for several τ values between 96 ns and 400 ns. A four-step phase cycle was used to remove unwanted echoes [75]. The raw time traces were baseline corrected with a third-order polynomial, apodized with a Hamming window, and zero filled. After the two-dimensional Fourier transformation, the absolute value spectra were calculated and plotted in 2D vs the two frequency axes.

Mims Electron Nuclear Double Resonance (ENDOR) [76]. ENDOR experiments were carried out with a Bruker ESP 380E spectrometer (X-band) equipped with an EN 4118X-MD4 Bruker resonator with the pulse sequence $\pi/2$-τ-$\pi/2$-T-$\pi/2$-τ-echo, with a $\pi/2$ pulse of 16 ns and a radiofrequency (rf) pulse of 10 μs length.

Simulations. CW-EPR and HYSCORE spectra were simulated using the EasySpin® toolbox for MATLAB [77]. The time traces obtained from the HYSCORE simulations were processed like the experimental ones, baseline corrected, apodised, and 2D-FFTransformed.

The simulations of the electron–nuclear spin system were performed using a spin Hamiltonian containing the electron Zeeman (EZ) term, a nuclear Zeeman (NZ), hyperfine (HF), and if $I > \frac{1}{2}$, a nuclear quadrupole (NQ) term for every magnetic nucleus interacting with the electron spin:

$$\hat{H}_0 = \hat{H}_{EZ} + \hat{H}_{NZ} + \hat{H}_{HF} + \hat{H}_{NQ} = \frac{2\pi \mu_B}{h} \vec{B}_0^T g \vec{\hat{S}} + \sum_i \vec{\hat{S}}^T \vec{A}_i \vec{\hat{I}}_i + \frac{2\pi \mu_N}{h} \sum_i g_{N,i} \vec{B}_0^T \vec{\hat{I}} + \sum_{I_i > \frac{1}{2}} \vec{\hat{I}}_i^T Q_i \vec{\hat{I}}_i \quad (2)$$

where μ_B is the Bohr magneton, h is the Planck constant, B_0 is the applied external magnetic field, $\vec{S}$ ($S = \frac{1}{2}$) and $\vec{I}$ are the electron and nuclear spin operators, and g, A, and Q are the gyromagnetic, hyperfine, and nuclear quadrupole tensors, respectively.

To have a complete Hamiltonian formulation, the summation i should be the overall magnetic nuclei coupled to the electron spin, four porphyrin ^{14}Ns, N-Im for the Im-inhibited protein, and several protons. However, to spare computational time, each nucleus was simulated separately and once the individual parameters were optimized, simulations considering more than one nucleus were performed selectively, specifically for those indicated in the text, to check for combination frequencies.

5. Conclusions

In this work, we presented a detailed analysis, by means of multifrequency Pulse EPR spectroscopy, of the active site of the peroxygenase-like CYP450, CYP116B5hd, in which the electronic location of the semi-occupied orbital is determined with respect to the heme site geometry and linked to the geometry of the axial ligands. Since during the reaction cycle, this protein receives 1 + 1 electrons from the reductase domain/partner that end up in the axial molecular oxygen ligand of the iron, the location of the orbitals where the transferred electron is hosted is a relevant piece of information to help understand the catalysis of this CYP450. The complete spin Hamiltonian parameters describing such a center were obtained with high accuracy and are consistent with similar systems previously characterized. The results showed that the imidazole binding, easily detectable with the joint effort of CW and Pulse EPR spectroscopy, does not severely alter the electronic environment of the Fe^{III}-heme system. Moreover, the inhibitor ring orientation can be obtained through its interaction with the unpaired electron. Since CYP450s are involved in the metabolism of xenobiotics such as anti-cancer drugs, the study of imidazole inhibition of CYP450 could be useful for the development of imidazole-containing drugs to be used as coadjuvants to prolong and increase the effect of cancer treatments.

Supplementary Materials: The following supporting information can be downloaded at: https://www.mdpi.com/article/10.3390/molecules29020518/s1, Figure S1: Experimental X-band CW-EPR spectra; Figure S2: X-band echo detected field sweep EPR; Figure S3: Dikanov Methodology for the analysis of 1H HYSCORE Spectra; Figure S4: Mims ENDOR spectra of CYP116B5hd in H$_2$O and D$_2$O; Figure S5: Q-band HYSCORE spectra of CYP116B5hd in H$_2$O—Experiments and simulations; Figure S6: HYSCORE spectra of CYP116B5hd interacting with imidazole-^{15}N$_2$; Figure S7: Complete HYSCORE spectra of CYP116B5hd in H2O; Figure S8: Complete HYSCORE spectra of CYP116B5hd in

D$_2$O; Figure S9: Complete HYSCORE spectra of CYP116B5hd interacting with imidazole in H$_2$O; and Figure S10: Complete HYSCORE spectra of CYP116B5hd interacting with imidazole-^{15}N$_2$ in H$_2$O.

Author Contributions: Investigation: A.F., G.M. and I.G.-R.; validation: A.F. and I.G.-R.; formal analysis: A.F., I.G.-R. and G.M.; writing—original draft: A.F. and I.G.-R.; data curation: A.F.; resources: D.C., M.C., I.G.-R., G.M. and G.D.N.; conceptualization: G.G., M.C. and I.G.-R.; writing—review and editing: D.C., G.D.N., G.G. and G.M.; funding acquisition: M.C. and I.G.-R.; and supervision, M.C. and I.G.-R. All authors have read and agreed to the published version of the manuscript.

Funding: This research was funded by the European Union's Horizon 2020 research and innovation programme under the Marie Skłodowska-Curie grant agreement No. 813209 and by grant No. PID2021-127287NB-I00 from the Spanish Ministry of Science and Innovation.

Institutional Review Board Statement: Not applicable.

Informed Consent Statement: Not applicable.

Data Availability Statement: Data can be found at https://zenodo.org/records/10535347 and https://zenodo.org/communities/paracat_community, accessed on 18 January 2024.

Acknowledgments: The authors would like to acknowledge the use of Servicio General de Apoyo a la Investigación-SAI, Universidad de Zaragoza.

Conflicts of Interest: The authors declare no conflicts of interest.

References

1. Ortiz de Montellano, P.R. *Cytochrome P450: Structure, Mechanism, and Biochemistry*, 4th ed.; Ortiz de Montellano, P.R., Ed.; Springer International Publishing: Cham, Switzerland, 2015; ISBN 978-3-319-12107-9.
2. Lamb, D.C.; Lei, L.; Warrilow, A.G.S.; Lepesheva, G.I.; Mullins, J.G.L.; Waterman, M.R.; Kelly, S.L. The First Virally Encoded Cytochrome P450. *J. Virol.* **2009**, *83*, 8266–8269. [CrossRef] [PubMed]
3. Nelson, D.R. Cytochrome P450 Diversity in the Tree of Life. *Biochim. Biophys. Acta (BBA) Proteins Proteom.* **2018**, *1866*, 141–154. [CrossRef] [PubMed]
4. Hammer, S.C.; Kubik, G.; Watkins, E.; Huang, S.; Minges, H.; Arnold, F.H. Anti-Markovnikov Alkene Oxidation by Metal-Oxo–Mediated Enzyme Catalysis. *Science (1979)* **2017**, *358*, 215–218. [CrossRef] [PubMed]
5. Tavanti, M.; Porter, J.L.; Sabatini, S.; Turner, N.J.; Flitsch, S.L. Panel of New Thermostable CYP116B Self-Sufficient Cytochrome P450 Monooxygenases That Catalyze C–H Activation with a Diverse Substrate Scope. *ChemCatChem* **2018**, *10*, 1042–1051. [CrossRef]
6. Correddu, D.; Di Nardo, G.; Gilardi, G. Self-Sufficient Class VII Cytochromes P450: From Full-Length Structure to Synthetic Biology Applications. *Trends Biotechnol.* **2021**, *39*, 1184–1207. [CrossRef]
7. Carta, M.; Malpass-Evans, R.; Croad, M.; Rogan, Y.; Jansen, J.C.; Bernardo, P.; Bazzarelli, F.; McKeown, N.B. An Efficient Polymer Molecular Sieve for Membrane Gas Separations. *Science (1979)* **2013**, *339*, 303–307. [CrossRef] [PubMed]
8. Farwell, C.C.; McIntosh, J.A.; Hyster, T.K.; Wang, Z.J.; Arnold, F.H. Enantioselective Imidation of Sulfides via Enzyme-Catalyzed Intermolecular Nitrogen-Atom Transfer. *J. Am. Chem. Soc.* **2014**, *136*, 8766–8771. [CrossRef]
9. Farwell, C.C.; Zhang, R.K.; McIntosh, J.A.; Hyster, T.K.; Arnold, F.H. Enantioselective Enzyme-Catalyzed Aziridination Enabled by Active-Site Evolution of a Cytochrome P450. *ACS Cent. Sci.* **2015**, *1*, 89–93. [CrossRef]
10. Zhang, X.; Li, S. Expansion of Chemical Space for Natural Products by Uncommon P450 Reactions. *Nat. Prod. Rep.* **2017**, *34*, 1061–1089. [CrossRef]
11. Prier, C.K.; Zhang, R.K.; Buller, A.R.; Brinkmann-Chen, S.; Arnold, F.H. Enantioselective, Intermolecular Benzylic C–H Amination Catalysed by an Engineered Iron-Haem Enzyme. *Nat. Chem.* **2017**, *9*, 629–634. [CrossRef]
12. Correddu, D.; Helmy Aly, S.; Di Nardo, G.; Catucci, G.; Prandi, C.; Blangetti, M.; Bellomo, C.; Bonometti, E.; Viscardi, G.; Gilardi, G. Enhanced and Specific Epoxidation Activity of P450 BM3 Mutants for the Production of High Value Terpene Derivatives. *RSC Adv.* **2022**, *12*, 33964–33969. [CrossRef] [PubMed]
13. Rodriguez-Antona, C.; Ingelman-Sundberg, M. Cytochrome P450 Pharmacogenetics and Cancer. *Oncogene* **2006**, *25*, 1679–1691. [CrossRef] [PubMed]
14. Pearson, J.; Dahal, U.P.; Rock, D.; Peng, C.-C.; Schenk, J.O.; Joswig-Jones, C.; Jones, J.P. The Kinetic Mechanism for Cytochrome P450 Metabolism of Type II Binding Compounds: Evidence Supporting Direct Reduction. *Arch. Biochem. Biophys.* **2011**, *511*, 69–79. [CrossRef]
15. Deodhar, M.; Al Rihani, S.B.; Arwood, M.J.; Darakjian, L.; Dow, P.; Turgeon, J.; Michaud, V. Mechanisms of CYP450 Inhibition: Understanding Drug-Drug Interactions Due to Mechanism-Based Inhibition in Clinical Practice. *Pharmaceutics* **2020**, *12*, 846. [CrossRef] [PubMed]
16. Guengerich, F.P. Inhibition of Cytochrome P450 Enzymes by Drugs-Molecular Basis and Practical Applications. *Biomol. Ther. (Seoul)* **2022**, *30*, 1–18. [CrossRef] [PubMed]

17. Minerdi, D.; Sadeghi, S.J.; Di Nardo, G.; Rua, F.; Castrignanò, S.; Allegra, P.; Gilardi, G. CYP116B5: A New Class VII Catalytically Self-sufficient Cytochrome P450 from *Acinetobacter radioresistens* That Enables Growth on Alkanes. *Mol. Microbiol.* **2015**, *95*, 539–554. [CrossRef] [PubMed]
18. Eser, B.E.; Zhang, Y.; Zong, L.; Guo, Z. Self-Sufficient Cytochrome P450s and Their Potential Applications in Biotechnology. *Chin. J. Chem. Eng.* **2021**, *30*, 121–135. [CrossRef]
19. Ciaramella, A.; Catucci, G.; Di Nardo, G.; Sadeghi, S.J.; Gilardi, G. Peroxide-Driven Catalysis of the Heme Domain of *A. radioresistens* Cytochrome P450 116B5 for Sustainable Aromatic Rings Oxidation and Drug Metabolites Production. *New Biotechnol.* **2020**, *54*, 71–79. [CrossRef]
20. Groves, J.T. Key Elements of the Chemistry of Cytochrome P-450: The Oxygen Rebound Mechanism. *J. Chem. Educ.* **1985**, *62*, 928. [CrossRef]
21. Sarkar, M.R.; Houston, S.D.; Savage, G.P.; Williams, C.M.; Krenske, E.H.; Bell, S.G.; De Voss, J.J. Rearrangement-Free Hydroxylation of Methylcubanes by a Cytochrome P450: The Case for Dynamical Coupling of C-H Abstraction and Rebound. *J. Am. Chem. Soc.* **2019**, *141*, 19688–19699. [CrossRef]
22. Correddu, D.; Catucci, G.; Giuriato, D.; Di Nardo, G.; Ciaramella, A.; Gilardi, G. Catalytically Self-sufficient CYP116B5: Domain Switch for Improved Peroxygenase Activity. *Biotechnol. J.* **2023**, *18*, 2200622. [CrossRef]
23. Munro, A.W.; Leys, D.G.; McLean, K.J.; Marshall, K.R.; Ost, T.W.B.; Daff, S.; Miles, C.S.; Chapman, S.K.; Lysek, D.A.; Moser, C.C.; et al. P450 BM3: The Very Model of a Modern Flavocytochrome. *Trends Biochem. Sci.* **2002**, *27*, 250–257. [CrossRef] [PubMed]
24. Peisach, J.; Blumberg, W.E. Electron Paramagnetic Resonance Study of the High- and Low-Spin Forms of Cytochrome P-450 in Liver and in Liver Microsomes from a Methylcholanthrene-Treated Rabbit. *Proc. Natl. Acad. Sci. USA* **1970**, *67*, 172–179. [CrossRef] [PubMed]
25. Lipscomb, J.D. Electron Paramagnetic Resonance Detectable States of Cytochrome P-450cam. *Biochemistry* **1980**, *19*, 3590–3599. [CrossRef] [PubMed]
26. Davydov, R.; Matsui, T.; Fujii, H.; Ikeda-Saito, M.; Hoffman, B.M. Kinetic Isotope Effects on the Rate-Limiting Step of Heme Oxygenase Catalysis Indicate Concerted Proton Transfer/Heme Hydroxylation. *J. Am. Chem. Soc.* **2003**, *125*, 16208–16209. [CrossRef]
27. Aldag, C.; Gromov, I.A.; García-Rubio, I.; von Koenig, K.; Schlichting, I.; Jaun, B.; Hilvert, D. Probing the Role of the Proximal Heme Ligand in Cytochrome P450cam by Recombinant Incorporation of Selenocysteine. *Proc. Natl. Acad. Sci. USA* **2009**, *106*, 5481–5486. [CrossRef]
28. Rittle, J.; Green, M.T. Cytochrome P450 Compound I: Capture, Characterization, and C-H Bond Activation Kinetics. *Science (1979)* **2010**, *330*, 933–937. [CrossRef]
29. Maurelli, S.; Chiesa, M.; Giamello, E.; Di Nardo, G.; Ferrero, V.E.V.; Gilardi, G.; Van Doorslaer, S. Direct Spectroscopic Evidence for Binding of Anastrozole to the Iron Heme of Human Aromatase. Peering into the Mechanism of Aromatase Inhibition. *Chem. Commun.* **2011**, *47*, 10737. [CrossRef]
30. Lockart, M.M.; Rodriguez, C.A.; Atkins, W.M.; Bowman, M.K. CW EPR Parameters Reveal Cytochrome P450 Ligand Binding Modes. *J. Inorg. Biochem.* **2018**, *183*, 157–164. [CrossRef]
31. Greule, A.; Izoré, T.; Machell, D.; Hansen, M.H.; Schoppet, M.; De Voss, J.J.; Charkoudian, L.K.; Schittenhelm, R.B.; Harmer, J.R.; Cryle, M.J. The Cytochrome P450 OxyA from the Kistamicin Biosynthesis Cyclization Cascade Is Highly Sensitive to Oxidative Damage. *Front. Chem.* **2022**, *10*, 868240. [CrossRef]
32. Famulari, A.; Correddu, D.; Di Nardo, G.; Gilardi, G.; Chiesa, M.; García-Rubio, I. EPR Characterization of the Heme Domain of a Self-Sufficient Cytochrome P450 (CYP116B5). *J. Inorg. Biochem.* **2022**, *231*, 111785. [CrossRef] [PubMed]
33. Famulari, A.; Correddu, D.; Di Nardo, G.; Gilardi, G.; Chiesa, M.; García-Rubio, I. CYP116B5hd, a Self-Sufficient P450 Cytochrome: A Dataset of Its Electronic and Geometrical Properties. *Data Brief* **2022**, *42*, 108195. [CrossRef] [PubMed]
34. Taylor, C.P.S. The EPR of Low Spin Heme Complexes Relation of the T2g Hole Model to the Directional Properties of the g Tensor, and a New Method for Calculating the Ligand Field Parameters. *Biochim. Biophys. Acta (BBA)—Protein Struct.* **1977**, *491*, 137–148. [CrossRef]
35. Alonso, P.; Martinez, J.; García-Rubio, I. The Study of the Ground State Kramers Doublet of Low-Spin Heminic System Revisited. A Comprehensive Description of the EPR and Mössbauer Spectra. *Coord. Chem. Rev.* **2007**, *251*, 12–24. [CrossRef]
36. García-Rubio, I.; Martínez, J.I.; Picorel, R.; Yruela, I.; Alonso, P.J. HYSCORE Spectroscopy in the Cytochrome b_{559} of the Photosystem II Reaction Center. *J. Am. Chem. Soc.* **2003**, *125*, 15846–15854. [CrossRef] [PubMed]
37. Vinck, E.; Van Doorslaer, S.; Dewilde, S.; Mitrikas, G.; Schweiger, A.; Moens, L. Analyzing Heme Proteins Using EPR Techniques: The Heme-Pocket Structure of Ferric Mouse Neuroglobin. *JBIC J. Biol. Inorg. Chem.* **2006**, *11*, 467–475. [CrossRef]
38. García-Rubio, I.; Alonso, P.J.; Medina, M.; Martínez, J.I. Hyperfine Correlation Spectroscopy and Electron Spin Echo Envelope Modulation Spectroscopy Study of the Two Coexisting Forms of the Hemeprotein Cytochrome c_6 from *Anabaena* PCC 7119. *Biophys. J.* **2009**, *96*, 141–152. [CrossRef]
39. García-Rubio, I.; Mitrikas, G. Structure and Spin Density of Ferric Low-Spin Heme Complexes Determined with High-Resolution ESEEM Experiments at 35 GHz. *JBIC J. Biol. Inorg. Chem.* **2010**, *15*, 929–941. [CrossRef]
40. Conner, K.P.; Vennam, P.; Woods, C.M.; Krzyaniak, M.D.; Bowman, M.K.; Atkins, W.M. 1,2,3-Triazole–Heme Interactions in Cytochrome P450: Functionally Competent Triazole–Water–Heme Complexes. *Biochemistry* **2012**, *51*, 6441–6457. [CrossRef]

41. Podgorski, M.N.; Harbort, J.S.; Coleman, T.; Stok, J.E.; Yorke, J.A.; Wong, L.-L.; Bruning, J.B.; Bernhardt, P.V.; De Voss, J.J.; Harmer, J.R.; et al. Biophysical Techniques for Distinguishing Ligand Binding Modes in Cytochrome P450 Monooxygenases. *Biochemistry* **2020**, *59*, 1038–1050. [CrossRef]
42. Loftfield, R.B.; Eigner, E.A.; Pastuszyn, A.; Lövgren, T.N.; Jakubowski, H. Conformational Changes during Enzyme Catalysis: Role of Water in the Transition State. *Proc. Natl. Acad. Sci. USA* **1980**, *77*, 3374–3378. [CrossRef] [PubMed]
43. Meyer, E. Internal Water Molecules and H-bonding in Biological Macromolecules: A Review of Structural Features with Functional Implications. *Protein Sci.* **1992**, *1*, 1543–1562. [CrossRef] [PubMed]
44. Thellamurege, N.; Hirao, H. Water Complexes of Cytochrome P450: Insights from Energy Decomposition Analysis. *Molecules* **2013**, *18*, 6782–6791. [CrossRef] [PubMed]
45. Shaik, S.; Dubey, K.D. The Catalytic Cycle of Cytochrome P450: A Fascinating Choreography. *Trends Chem.* **2021**, *3*, 1027–1044. [CrossRef]
46. Yadav, S.; Kardam, V.; Tripathi, A.; T G, S.; Dubey, K.D. The Performance of Different Water Models on the Structure and Function of Cytochrome P450 Enzymes. *J. Chem. Inf. Model.* **2022**, *62*, 6679–6690. [CrossRef] [PubMed]
47. Thomann, H.; Bernardo, M.; Goldfarb, D.; Kroneck, P.M.H.; Ullrich, V. Evidence for Water Binding to the Fe Center in Cytochrome P450cam Obtained by 17O Electron Spin-Echo Envelope Modulation Spectroscopy. *J. Am. Chem. Soc.* **1995**, *117*, 8243–8251. [CrossRef]
48. Schweiger, A.; Jeschke, G. *Principles of Pulse Electron Paramagnetic Resonance*; Oxford University Press: Oxford, UK, 2001; ISBN 0198506341.
49. Ciaramella, A.; Catucci, G.; Gilardi, G.; Di Nardo, G. Crystal Structure of Bacterial CYP116B5 Heme Domain: New Insights on Class VII P450s Structural Flexibility and Peroxygenase Activity. *Int. J. Biol. Macromol.* **2019**, *140*, 577–587. [CrossRef]
50. LoBrutto, R.; Scholes, C.P.; Wagner, G.C.; Gunsalus, I.C.; Debrunner, P.G. Electron Nuclear Double Resonance of Ferric Cytochrome P450CAM. *J. Am. Chem. Soc.* **1980**, *102*, 1167–1170. [CrossRef]
51. Scholes, C.P.; Falkowski, K.M.; Chen, S.; Bank, J. Electron Nuclear Double Resonance (ENDOR) of Bis(Imidazole) Ligated Low-Spin Ferric Heme Systems. *J. Am. Chem. Soc.* **1986**, *108*, 1660–1671. [CrossRef]
52. Goldfarb, D.; Bernardo, M.; Thomann, H.; Kroneck, P.M.H.; Ullrich, V. Study of Water Binding to Low-Spin Fe(III) in Cytochrome P450 by Pulsed ENDOR and Four-Pulse ESEEM Spectroscopies. *J. Am. Chem. Soc.* **1996**, *118*, 2686–2693. [CrossRef]
53. García-Rubio, I.; Medina, M.; Cammack, R.; Alonso, P.J.; Martínez, J.I. CW-EPR and ENDOR Study of Cytochrome c_6 from Anabaena PCC 7119. *Biophys. J.* **2006**, *91*, 2250–2263. [CrossRef] [PubMed]
54. Davydov, R.; Makris, T.M.; Kofman, V.; Werst, D.E.; Sligar, S.G.; Hoffman, B.M. Hydroxylation of Camphor by Reduced Oxy-Cytochrome P450cam: Mechanistic Implications of EPR and ENDOR Studies of Catalytic Intermediates in Native and Mutant Enzymes. *J. Am. Chem. Soc.* **2001**, *123*, 1403–1415. [CrossRef] [PubMed]
55. Davydov, R.; Hoffman, B.M. Active Intermediates in Heme Monooxygenase Reactions as Revealed by Cryoreduction/Annealing, EPR/ENDOR Studies. *Arch. Biochem. Biophys.* **2011**, *507*, 36–43. [CrossRef] [PubMed]
56. Davydov, R.; Dawson, J.H.; Perera, R.; Hoffman, B.M. The Use of Deuterated Camphor as a Substrate in ^{1}H ENDOR Studies of Hydroxylation by Cryoreduced Oxy P450cam Provides New Evidence of the Involvement of Compound I. *Biochemistry* **2013**, *52*, 667–671. [CrossRef] [PubMed]
57. Davydov, R.; Im, S.; Shanmugam, M.; Gunderson, W.A.; Pearl, N.M.; Hoffman, B.M.; Waskell, L. Role of the Proximal Cysteine Hydrogen Bonding Interaction in Cytochrome P450 2B4 Studied by Cryoreduction, Electron Paramagnetic Resonance, and Electron–Nuclear Double Resonance Spectroscopy. *Biochemistry* **2016**, *55*, 869–883. [CrossRef] [PubMed]
58. Dikanov, S.A.; Bowman, M.K. Cross-Peak Lineshape of Two-Dimensional ESEEM Spectra in Disordered $S = 1/2, I = 1/2$ Spin Systems. *J. Magn. Reson. A* **1995**, *116*, 125–128. [CrossRef]
59. Dikanov, S.A.; Tyryshkin, A.M.; Bowman, M.K. Intensity of Cross-Peaks in Hyscore Spectra of $S = 1/2, I = 1/2$ Spin Systems. *J. Magn. Reson.* **2000**, *144*, 228–242. [CrossRef]
60. Verras, A.; Alian, A.; Montellano, P.R.O.D. Cytochrome P450 Active Site Plasticity: Attenuation of Imidazole Binding in Cytochrome P450cam by an L244A Mutation. *Protein Eng. Des. Sel.* **2006**, *19*, 491–496. [CrossRef]
61. Podgorski, M.N.; Harbort, J.S.; Lee, J.H.Z.; Nguyen, G.T.H.; Bruning, J.B.; Donald, W.A.; Bernhardt, P.V.; Harmer, J.R.; Bell, S.G. An Altered Heme Environment in an Engineered Cytochrome P450 Enzyme Enables the Switch from Monooxygenase to Peroxygenase Activity. *ACS Catal.* **2022**, *12*, 1614–1625. [CrossRef]
62. Vinck, E.; Van Doorslaer, S. Analysing Low-Spin Ferric Complexes Using Pulse EPR Techniques: A Structure Determination of Bis (4-Methylimidazole)(Tetraphenylporphyrinato)Iron(Iii). *Phys. Chem. Chem. Phys.* **2004**, *6*, 5324. [CrossRef]
63. Ashby, C.I.H.; Cheng, C.P.; Brown, T.L. Nitrogen-14 Nuclear Quadrupole Resonance Spectra of Coordinated Imidazole. *J. Am. Chem. Soc.* **1978**, *100*, 6057–6063. [CrossRef]
64. Griffith, J.S. Theory of E.P.R. in Low-Spin Ferric Haemoproteins. *Mol. Phys.* **1971**, *21*, 135–139. [CrossRef]
65. Zoppellaro, G.; Bren, K.L.; Ensign, A.A.; Harbitz, E.; Kaur, R.; Hersleth, H.; Ryde, U.; Hederstedt, L.; Andersson, K.K. Review: Studies of Ferric Heme Proteins with Highly Anisotropic/Highly Axial Low Spin (S = 1/2) Electron Paramagnetic Resonance Signals with Bis-Histidine and Histidine-methionine Axial Iron Coordination. *Biopolymers* **2009**, *91*, 1064–1082. [CrossRef] [PubMed]

66. Conner, K.P.; Cruce, A.A.; Krzyaniak, M.D.; Schimpf, A.M.; Frank, D.J.; Ortiz de Montellano, P.; Atkins, W.M.; Bowman, M.K. Drug Modulation of Water–Heme Interactions in Low-Spin P450 Complexes of CYP2C9d and CYP125A1. *Biochemistry* **2015**, *54*, 1198–1207. [CrossRef] [PubMed]
67. Le Breton, N.; Wright, J.J.; Jones, A.J.Y.; Salvadori, E.; Bridges, H.R.; Hirst, J.; Roessler, M.M. Using Hyperfine Electron Paramagnetic Resonance Spectroscopy to Define the Proton-Coupled Electron Transfer Reaction at Fe–S Cluster N2 in Respiratory Complex I. *J. Am. Chem. Soc.* **2017**, *139*, 16319–16326. [CrossRef]
68. Kondo, H.X.; Kanematsu, Y.; Masumoto, G.; Takano, Y. PyDISH: Database and Analysis Tools for Heme Porphyrin Distortion in Heme Proteins. *Database* **2023**, *2023*, baaa066. [CrossRef] [PubMed]
69. Peisach, J.; Mims, W.B.; Davis, J.L. Studies of the Electron-Nuclear Coupling between Fe(III) and 14N in Cytochrome P-450 and in a Series of Low Spin Heme Compounds. *J. Biol. Chem.* **1979**, *254*, 12379–12389. [CrossRef]
70. Magliozzo, R.S.; Peisach, J. Evaluation of Nitrogen Nuclear Hyperfine and Quadrupole Coupling Parameters for the Proximal Imidazole in Myoglobin-Azide, -Cyanide, and -Mercaptoethanol Complexes by Electron Spin Echo Envelope Modulation Spectroscopy. *Biochemistry* **1993**, *32*, 8446–8456. [CrossRef]
71. Garcia, M.L.S.; Smith, J.A.S.; Bavin, P.M.G.; Ganellin, C.R. ^{14}N and ^{2}H Quadrupole Double Resonance in Substituted Imidazoles. *J. Chem. Soc. Perkin Trans.* **1983**, *2*, 1391–1399. [CrossRef]
72. Hsieh, Y.-N.; Rubenacker, G.V.; Cheng, C.P.; Brown, T.L. Nitrogen-14 Nuclear Quadrupole Resonance Spectra of Coordinated Pyridine. *J. Am. Chem. Soc.* **1977**, *99*, 1384–1389. [CrossRef]
73. Omura, T.; Sato, R. The Carbon Monoxide-Binding Pigment of Liver Microsomes. II. Solubilization, Purification, and Properties. *J. Biol. Chem.* **1964**, *239*, 2379–2385. [CrossRef] [PubMed]
74. Höfer, P.; Grupp, A.; Nebenführ, H.; Mehring, M. Hyperfine Sublevel Correlation (Hyscore) Spectroscopy: A 2D ESR Investigation of the Squaric Acid Radical. *Chem. Phys. Lett.* **1986**, *132*, 279–282. [CrossRef]
75. Bodenhausen, G.; Kogler, H.; Ernst, R.R. Selection of Coherence-Transfer Pathways in NMR Pulse Experiments. *J. Magn. Reson. (1969)* **1984**, *58*, 370–388. [CrossRef]
76. Mims, W.B. Pulsed Endor Experiments. *Proc. R. Soc. Lond. A Math. Phys. Sci.* **1965**, *283*, 452–457. [CrossRef]
77. Stoll, S.; Schweiger, A. EasySpin, a Comprehensive Software Package for Spectral Simulation and Analysis in EPR. *J. Magn. Reson.* **2006**, *178*, 42–55. [CrossRef]

Disclaimer/Publisher's Note: The statements, opinions and data contained in all publications are solely those of the individual author(s) and contributor(s) and not of MDPI and/or the editor(s). MDPI and/or the editor(s) disclaim responsibility for any injury to people or property resulting from any ideas, methods, instructions or products referred to in the content.

Article

Dating Sediments by EPR Using Al-h Centre: A Comparison between the Properties of Fine (4–11 µm) and Coarse (>63 µm) Quartz Grains

Zuzanna Kabacińska [1,*] and Alida Timar-Gabor [1,2]

1. Interdisciplinary Research Institute on Bio-Nano-Sciences, Babeș-Bolyai University, Treboniu Laurian 42, 400271 Cluj-Napoca, Romania; alida.timar@ubbcluj.ro
2. Faculty of Environmental Science and Engineering, Babeș-Bolyai University, Fântânele 30, 400000 Cluj-Napoca, Romania
* Correspondence: zuzanna.kabacinska@ubbcluj.ro

Citation: Kabacińska, Z.; Timar-Gabor, A. Dating Sediments by EPR Using Al-h Centre: A Comparison between the Properties of Fine (4–11 µm) and Coarse (>63 µm) Quartz Grains. *Molecules* 2022, 27, 2683. https://doi.org/10.3390/molecules27092683

Academic Editors: Yordanka Karakirova and Nicola D. Yordanov

Received: 30 March 2022
Accepted: 19 April 2022
Published: 21 April 2022

Publisher's Note: MDPI stays neutral with regard to jurisdictional claims in published maps and institutional affiliations.

Copyright: © 2022 by the authors. Licensee MDPI, Basel, Switzerland. This article is an open access article distributed under the terms and conditions of the Creative Commons Attribution (CC BY) license (https://creativecommons.org/licenses/by/4.0/).

Abstract: The possibility of EPR dating for sediments using Al-h signals of fine (4–11 µm) grains of quartz has not been previously discussed. Here, the Al-h and peroxy EPR spectra of fine (4–11 µm) and coarse (63–90, 125–180 µm) sedimentary quartz from thoroughly investigated loess sites in Eastern Europe were examined. By comparing experimental spectra with a simulated signal, we evaluated the overestimation observed when using the standard approach established by Toyoda and Falguères to measure Al-h intensity for different doses of radiation, up to 40,000 Gy. This overestimation, caused by the presence of peroxy signals, was much more pronounced for fine grains. Fine grains exhibited some additional dose-dependent signals, which, for some samples, caused a complete distortion of the Al-h spectra at high doses, making it impossible to measure the standard amplitude. We propose a new approach to measuring Al-h signal intensity, focusing on the peak-to-baseline amplitude of the part of the signal at g ≈ 2.0603, which is not affected by the peroxy signals and therefore has the potential of providing more accurate results. The shapes of dose response curves constructed for coarse and fine grains using the new approach show considerable similarity, suggesting that Al-h centre formation in fine and coarse grains upon artificial radiation at room temperature follows the same pattern.

Keywords: electron paramagnetic resonance (EPR); electron spin resonance (ESR); quartz; Al-h centre; fine grains; dose response curve

1. Introduction

Quartz (SiO_2) is a material of great importance in many areas of Earth sciences, as well as in industry. As all crystals, it contains a vast number of point defects, which may be either intrinsic (involving only atoms of the host lattice—vacancies, interstitial atoms and excess atoms) or extrinsic (belonging to foreign atoms in lattice and inter-lattice positions) [1,2]. Those of most interest in the field of geochronology include Si- and O-vacancies and impurity related defects. Among the latter, Al^{3+} always presents in quartz, substituting for Si^{4+} with charge compensation generally achieved by Li^+, Na^+ or H^+, which gives rise to $[AlO_4/M^+]^0$ (where M^+ denotes an alkali metal or hydrogen ion) [3]. Ti^{4+} may substitute for Si^{4+} in quartz with no charge compensation, creating $[TiO_4]^0$ [4]. Ge centre, namely $[GeO_4/M^+]^0$ (most notably $[GeO_4/Li^+]^0$) is sometimes observed in irradiated natural quartz [4,5]. A neutral oxygen vacancy can trap an electronic hole, forming a paramagnetic oxygen vacancy (E_1' centre) [4,5]. Performing systematic investigations on quartz using electron paramagnetic resonance (EPR) (or electron spin resonance—ESR) spectroscopy, a method of high sensitivity, allows for gaining a deeper understanding of the mechanisms involved when the defects in quartz are subjected to irradiation.

EPR has been applied in dating geological and archaeological materials for over 40 years. Together with optically stimulated luminescence (OSL) and thermoluminescence (TL), the so-called trapped-charge dating methods have been extensively used for dating sediments using quartz (e.g., [6–9]). Quartz records the amount of ionising radiation it has been exposed to as a latent signal within its crystal lattice, and therefore can be used as a natural dosimeter for quantifying the radiation history of materials. Irradiation at room temperature leads to the dissociation of $[AlO_4/M^+]^0$, resulting in the formation of $[AlO_4/H^+]^0$ and $[AlO_4]^0$ [10]. $[AlO_4]^0$, also referred to as Al-hole or Al-h, as a paramagnetic centre, is therefore detectable by EPR, and has been extensively used for dating sediments [6,7,9,11–19]. Ti centres have also been widely used for dating [6,11,18–20]. Upon room temperature irradiation, Ti^{4+} may trap an electron together with an alkali ion M^+ for charge compensation, forming $[TiO_4/M^+]^0$, where M^+ can be either Li^+, H^+ or Na^+ [4]. Trapped-charge dating methods are based on the assumption that the natural growth of the signal of interest can be reproduced by laboratory irradiation, which leads to the construction of a dose response curve—a plot of EPR intensity versus the doses of irradiation, obtained separately for every investigated sample. The equivalent dose—a total dose of radiation absorbed by the crystal, giving rise to the signal measured in the natural sample—is determined by extrapolation (in the case of additive dose protocols) or interpolation (in the case of regenerative protocols) of the dose response curve.

In many luminescence and EPR dating studies, the choice of grain size fraction used for analysis has been most often dictated predominantly by the nature and availability of the material. Based on a series of previous research, Timar-Gabor et al. [21] showed that there is a discrepancy in the ages obtained by the single aliquot regeneration protocol (SAR) OSL between different grain sizes and an age underestimation for finer grains, and suggested a potentially worldwide phenomenon. However, defects giving rise to luminescence in quartz have not yet been unambiguously identified, and their correlation with the defects detected by EPR remains unestablished. Consequently, observations concerning grain size effects based on luminesce results are not directly transferable to EPR defects, which leaves this topic largely unexplored.

To fill this gap, a systematic approach needs to be employed, starting with a thorough investigation of the dependence of EPR intensity of defects on grain size, and followed by experiments showing their behaviour when subjected to laboratory irradiation, which would expose any possible differences between fine- and coarse-grained quartz. It goes without saying that such experimental studies should also be complemented by the development of appropriate models. The first objective was addressed by Timar-Gabor [22] by showing the dependence of EPR intensity of the main paramagnetic defects in quartz with grain size, for fraction 4–11, 63–90, 90–125, 125–180, and 180–250 µm. The intensity of the $E_1{}'$ and Al-h signal in natural samples was found to decrease with increasing grain size, while $[TiO_4/Li^+]^0$ signals, detected only in coarse fractions, increased with increasing grain size. The second objective, the investigation the behaviour of the defects under laboratory irradiation, is the subject of this work. To achieve any of these goals, or, in fact, to obtain any reliable dating, an accurate measurement of EPR signal intensity is crucial.

In this study, we focus on the Al-h signal in fine (4–11 µm) and coarse (>60 µm) grains. The Al-h signal can only be measured by EPR at cryogenic temperatures due to the very short spin–lattice relaxation time of the defect. It produces a complex EPR spectrum arising from the interaction of the unpaired electron with nearby magnetic nuclei. The Al-h signal consists of a central set of peaks around g ≈ 2.008 displaying a distinct hyperfine structure, and a much less intense set of peaks at about g ≈ 2.06. In early attempts at dating using the Al-h signal [23–25], different peaks from the central set were considered for evaluating its intensity. Their reliability was compared in a study by Lin et al. [26]. Eventually, a common approach proposed by Toyoda and Falguères [27] was adopted, and it has been widely used ever since by the EPR dating community (e.g., [6,9,15–17]). This approach is based on the measurement of a peak-to-peak amplitude between the top of the first peak (g = 2.018) and the bottom of the last peak (g = 1.993) of the central part of the [27]Al hyperfine structure. This method has been extremely useful due to its simplicity and the fact that it focuses on the most distinct peaks, which are clearly distinguishable even for very weak signals. However, its applicability is sometimes limited by the presence of additional signals superimposed on the central part of the Al-h signal.

These additional signals are referred to as "peroxy" species or, for simplicity, sometimes as a peroxy centre (singular), although their spectrum is clearly composed of many overlapping signals. They are visible most clearly at room temperature, when the Al-h is not detectable. Friebele et al. [28] first established a peroxy radical in neutron or gamma-ray irradiated ^{17}O-enriched fused silica and suggested it derives from pre-existing bridging peroxy linkages ($\equiv$Si–O–O–Si$\equiv$, where $\equiv$ represents three Si-O bonds), which shed an electron to form peroxy radicals by irradiation and/or thermal treatment. Peroxy radicals in crystalline SiO$_2$ (α-quartz) have been suggested by several EPR studies and established by Botis et al. [29,30], Nilges et al. [31,32] and Pan et al. [33,34] in their very detailed studies. Based on their research, it was concluded that most of the discrepancies in the literature concerning the g-factor values, linewidths and hyperfine structure reported for the peroxy centres can be attributed to incompletely resolved site splittings in previous X- and Q-band studies. For an in-depth investigation of these species, higher microwave frequencies should be applied, but the accessibility of such equipment is very limited. Despite the wealth of information provided by these studies, there are still many unanswered questions regarding the nature of these signals, answers to which are essential considering their relevance in EPR dating and provenance investigations. It should be noted that, apart from peroxy centres, another type of oxygen excess centre has been identified, namely, the non-bridging oxygen hole centre (NBOHC), described as oxygen dangling bonds $\equiv$Si–O· (where · represents an unpaired electron) [35]. For simplicity, however, in this study, we use the term "peroxy" to describe all signals observed between $g \approx 2.01$ and $g \approx 1.99$ at room temperature, with the exception of E' and Ge centres.

The complexity of the peroxy spectrum, combined with the limitations of the X-band spectroscopy routinely used for dating, makes attempts at isolating these signals to obtain an undisturbed signal from the Al-h centre extremely challenging. Perhaps for this reason, the issue of peroxy signals interfering with Al-h measurements has been largely ignored in the literature. However, some amendments have been occasionally employed to circumvent the problem, such as subtracting the overlapping peroxy signal intensity using its EPR signal intensity after annealing [36]. The assumption here is that the peroxy signal changes neither with heating nor with the dose of irradiation, and the same signal can be used for subtraction at low and high doses. While the latter assumption is generally accepted in the case of coarse grains, this has not been confirmed for fine grains. Indeed, Timar-Gabor [22] reported on a dose-dependent signal at $g \approx 2.011$, detected in a fine-grained fraction of quartz, which suggests that this approach might not be applicable in every case. Moreover, it introduces additional uncertainty related to the determination of peroxy signal intensity. Another approach, used by Tsukamoto et al. [7] for Al-h measurements conducted at 123 K, when the ^{27}Al hyperfine structures were not visible, was based on the measurement of the peak-to-peak intensity of the first central peak. It was reported to be consistent with the peak-to-peak intensity of the whole peak minus the intensity of the peroxy centre, which was measured at 183 K. No significant changes in the peroxy centre intensity were observed when raising the temperature from 123 to 183 K, and the Al-h signal at 183 K became almost undetectable. As in the previous example, this method bears some additional uncertainties. Additionally, the authors noted that their measurements might not be directly comparable with other studies, which use the traditional approach. It is clear that developing an alternative approach to measuring Al-h signal intensity, unaffected by the presence of the peroxy signals, would improve the accuracy of age determination and, therefore, greatly benefit the EPR dating community.

The aforementioned study by Timar-Gabor [22], conducted on several samples, including two (ROX 1.14 and STY 1.10) studied in this work, shows that peroxy signals have significantly higher intensities in fine grains (4–11 µm) and decrease when grain size increases. Extended etching experiments resulted in obtaining partial evidence that these defects are concentrated in damaged areas of the grains. The weaker signals of peroxy centres would suggest that coarser fractions should be preferred for conventional EPR dating using the approach of Toyoda and Falguères [27] to measure Al-h intensity. However, assuming that the issue of accurately determining Al-h signal intensity could be solved, finer grains would not have to be automatically dismissed solely on that basis, especially as they are the main constituents of many sedimentary archives, such as loess, lake or marine sediments. That would allow for a thorough comparison of the properties of the Al-h signal observed in fine and coarse grains and open the possibility of EPR dating based on fine grains, which has not been discussed before.

In this work, we propose an alternative method of evaluating Al-h signal intensity, which circumvents the issue of interfering peroxy signals. The results obtained for the new approach and the standard approach by Toyoda and Falguères [27] are compared using the measurements of fine (4–11 μm) and coarse (63–90, 125–180 μm) quartz separates from thoroughly investigated loess palaeosol sites in Eastern Europe (Roxolany, Stayky and Mircea Vodă), which were used in the previous investigations carried out by our group. By comparing the experimental spectra with a simulated signal of the Al-h centre, we evaluated the overestimation that results from using the standard approach for different doses of radiation, up to 40,000 Gy. We then used the dose response curves constructed from the intensities obtained with the new approach to compare, for the first time, the response of the Al-h signal to laboratory irradiation displayed by fine- and coarse-grained fractions.

2. Materials and Methods

2.1. Samples

Experiments presented in this study were conducted on archived quartz separates of different grain sizes from previous investigations carried out by our group.

Sample Rox 1.14 originates from Roxolany, loess palaeosol section, Southern Ukraine, and was collected below the Brunhes/Matuyama polarity transition. The results of EPR dating using a multicentre approach, along with optically stimulated investigations using both the standard single aliquot regenerative (SAR) multigrain OSL procedure, as well as single grain investigations, are presented in detail in [37].

Quartz sample Sty 1.10 comes from Stayky, loess palaeosol section, Northern Ukraine. The OSL chronology of this section, as well as extended SAR-OSL dose response curves on the Styky samples, are presented in detail in [38].

Sample 2 MV 80 was collected near the village of Mircea Vodă, which is situated in the Dobrogea plateau of SE Romania, about 15 km from the Danube River. Optical dating results for this site were published in [39] (including this sample) and in [8,40] (previous sampling).

Preparation protocol, following the standard OSL preparation guidelines, is described in detail in the aforementioned references, as well as in [22].

The selection of samples for the current study was based on the availability of sufficient material for EPR investigation and the high purity of the quartz extracts, as confirmed by routine tests in OSL dating as well as by scanning electron microscopy imaging, coupled with energy dispersive X-ray spectroscopy (EDX) [22].

2.2. EPR Measurements

EPR measurements were performed on an X-band Bruker EMX Plus spectrometer at Babeș-Bolyai University, Cluj-Napoca, Romania. All samples were placed in quartz glass tubes filled with a mass of 200 mg ± 10% for coarse grains (>63 μm) and 100 mg ± 10% for fine grains, maintaining the same volume, and with measurements later normalized to 100 mg for intercomparison. Care was taken that all samples were centred inside the cavity. Samples were rotated in the cavity using a programmable goniometer and measured at 3 different angles (every 120°, 1 scan per angle). Measurements were usually repeated 2–5 times at a few weeks' intervals. Details of reproducibility tests are described in [22].

A mean EPR intensity was used for constructing a dose response curve, and standard error was indicated in all the plots. Exposure of samples to sunlight during measurements was restricted to a minimum. Measurements were carried out at 90 K for Al-h centres and at room temperature (295 K) for peroxy centres, using a variable temperature unit. Spectra were acquired using the following settings: 3350 ± 150 G scanned magnetic field, modulation amplitude 1 G, modulation frequency 100 kHz, microwave power 2 mW, conversion time 40 ms, time constant 40 ms. Baseline correction was performed when necessary using Bruker's Xenon software.

Samples were gamma-irradiated with doses up to 10,000 Gy on top of the natural dose for sample ROX 1.14, and up to 40,000 Gy for samples STY 1.10 and 2 MV 80. Due to the limited availability of the material, fewer aliquots were obtained for the coarse fraction than for the fine fraction. Gamma irradiations were performed at room temperature at the Department of Health Technology at DTU (Dosimetry Research Unit) in Denmark using a calibrated ^{60}Co gamma cell with a dose rate of 2 Gy/s (dose rate to water) at the time of irradiation. Dose rate to quartz was estimated to be 96% of dose rate to water based on Monte Carlo simulation considering the irradiation geometry used, as in [22].

2.3. Al-h Signal Simulations

Al-h signal was simulated with EasySpin [41] using parameters listed in Table 1. Initial parameters were based on [5,42] and adjusted to fit the experimental spectra. The values of quadrupole splitting were used as in [42] with no adjustments. When comparing with an experimental spectrum, an average of baseline-corrected experimental spectra obtained for a given dose was used. A spectrum recorded at a microwave frequency of 9.42 GHz was chosen as a reference, and magnetic field values for all spectra were adjusted to match the position of the signal. A simulated spectrum is shown in Figure 1, together with the principal components of the g-tensor values mentioned in the text.

Table 1. Spin Hamiltonian parameters used for simulating $[AlO_4]^0$ spectrum. A—hyperfine splitting, Q—quadrupole splitting. S = 1/2, ^{27}Al (I = 5/2), Lorentzian peak-to-peak linewidth 0.185 mT.

Parameter		x	y	z
g-Tensor		2.0603	2.0083	2.0021
A	(MHz)	14	17	18.2
	(mT)	0.499	0.606	0.649
Q	(MHz)	−0.62	−0.43	1.05
	(mT)	−0.022	−0.015	0.037

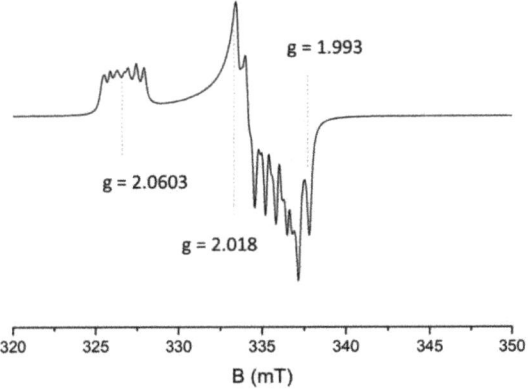

Figure 1. A simulated EPR signal of the Al-h centre with the g-factor values mentioned in the text. The parameters used for simulation are listed in Table 1.

3. Results and Discussion

3.1. Contribution of Peroxy Signals to Al-h Signal Measurements

The spectra of the Al-h and peroxy signals obtained for the coarse and fine grains of the quartz irradiated with different doses were compared based on three examples: sample ROX 1.14, STY 1.10 and 2 MV 80. Due to the different sources of quartz, one would expect these three samples to have different types and concentrations of defects, which makes them great subjects for studying the diversity of signals recorded by EPR.

3.1.1. Sample ROX 1.14

Figure 2 shows a comparison of coarse (125–180 µm) and fine (4–11 µm) quartz EPR spectra for sample ROX 1.14 acquired at 90 K and at room temperature. Both fractions exhibit clear differences in the shape of the spectra. Experimental spectra of natural and additionally irradiated (with 1000 and 10,000 Gy) samples recorded at 90 K were overlaid with a simulated spectrum of Al-h (Figure 2a,b). The shape of the experimental spectra differs from the simulated one, which is caused by overlapping with signals assigned to the so-called peroxy species. The difference between Al-h simulation and experimental spectra recorded at 90 K is much more significant in the case of fine grains, as the peroxy signals in 4–11 µm quartz are much stronger than in the bigger fractions, which was previously reported by Timar-Gabor [22]. In the case of coarse grains, this difference is visible only in the centre of the spectra and remains more pronounced for the smaller doses of irradiation.

The peroxy signals can be clearly registered at room temperature, when Al-h signal is not detectable (Figure 2c,d). The structure of the spectra is complex and consists of several overlapping signals. Their detailed characterisation and interpretation have been a subject of several studies (e.g., [29–34]) and is beyond the scope of this work. What is relevant for this study is whether the intensity of some of these signals depends on the dose of irradiation, an issue which, to our knowledge, has not been addressed in the literature. The only exception is a mention of a dose-dependent signal at $g \approx 2.011$ detected by Timar-Gabor [22] in fine grains. The spectra of coarse-grained quartz shown in Figure 2c indicates that only two signals, at $g \approx 2.000$ and $g \approx 1.996$, increase with the applied dose, while the rest do not show any changes. The signal at $g \approx 2.000$ can be ascribed to the E' centre and the peak at $g \approx 1.996$ to the Ge centre, namely, $[GeO_4/Li^+]^0$ [5] (their EPR spectra can be found therein). The peroxy signals detected in fine grains (Figure 2d) are much stronger, and the presence of some additional peaks is visible. In addition to the E' signal observed at $g \approx 1.999$ and the Ge signal at $g \approx 1.994$, at least two other signals, at $g \approx 2.009$ and $g \approx 2.001$, also show an increase with an increasing dose. The presence of these dose-dependent signals strongly influences the overall shape of the spectrum at high doses. The precise relationship between the intensity of these signals and the dose of laboratory irradiation cannot be determined at this point, as it requires separating them from the overlapping peaks, which is not possible without the aid of simulations and/or measurements at higher microwave frequencies.

As the peroxy signals in the 125–180 µm fraction of ROX 1.14 (and most of them in the case of the fine grains) are not dose-dependent, their contribution to the overall intensity of the signals registered in the considered range at low temperature decreases with the dose of radiation due to the increase in the Al-h signal. For coarse grains (Figure 2a), the experimental spectrum at high doses is very close in shape to the simulated one, while for fine grains (Figure 2b), the difference is still clearly visible.

When overlaying the experimental spectrum with a simulated one, we were faced with the issue of properly adjusting the amplitude of the latter. Since the central part of experimental spectra has proven to be distorted, to a varying degree, it should not be used as a reference point to adjust the simulated spectra. Therefore, a logical course of action was to choose peaks in the low-field part of the spectral range, specifically, the centre of the peak around $g \approx 2.0603$ (see Figure 1), and match the amplitude of the simulated spectrum to the experimental spectrum each time, using this point as a reference.

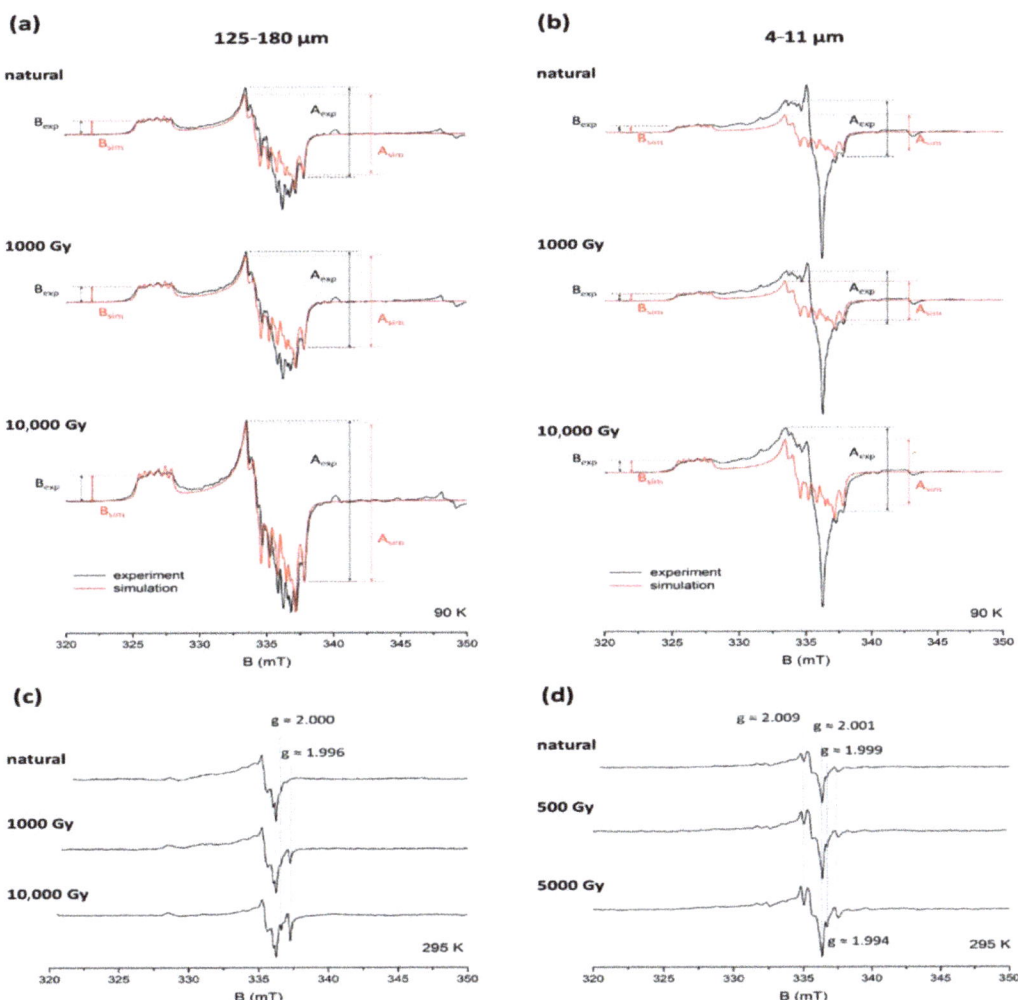

Figure 2. EPR spectra of the coarse (125–180 μm) and fine (4–11 μm) fraction of sample ROX 1.14, natural and additionally irradiated with 1000 and 10,000 Gy, recorded at 90 K ((**a**,**b**), respectively) and at room temperature ((**c**,**d**), respectively), and simulated spectra of Al-h signal (in red). Amplitudes A_{exp}, A_{sim}, B_{exp} and B_{sim} are marked with arrows. Major dose-dependent signals observed at room temperature are marked with blue dashed lines. The 90 K spectra for coarse grains are multiplied by a factor of 2.6.

The interference of peroxy signals may naturally cause problems for accurate measurements of Al-h signal intensity. A well-established method of measuring the intensity of the Al-h signal is based on the measurement of peak-to-peak amplitude between the top of the first peak of the central signal ($g = 2.018$) and the bottom of the last peak ($g = 1.993$) (see Figure 1) [27]. In Figure 2a,b we mark the amplitudes measured using this approach (denoted further as "A"), obtained from the experimental (A_{exp}) and simulated (A_{sim}) spectra of ROX 1.14 sample. As demonstrated for the additional dose of 10,000 Gy, both A_{exp} and A_{sim} give basically the same value (less than 2% difference) for coarse grains at high doses. However, due to the greater contribution of peroxy signals, A_{exp} amplitude at low doses is slightly overestimated compared to A_{sim} for the natural sample, giving about 13% and, for

1000 Gy, about 5% higher value compared to A_{sim}. For fine grains, the overestimation of A_{exp} compared to A_{sim} is much more significant. For the natural sample of 4–11 μm ROX 1.14, it amounts to approximately 54%; for natural + 1000 Gy, about 38%; and for natural + 10,000 Gy, about 27%, as a result of the increasing contribution of dose-dependent Al-h signals and the decreasing contribution of mostly non-dose-dependent peroxy signals.

It is therefore clear that, although the approach based on measuring amplitude A works very well for samples of coarse-grained quartz which have accumulated a high dose of irradiation (e.g., very old samples and laboratory-irradiated samples), it can result in a significant overestimation in the case of fine-grained and young coarse-grained quartz, which can affect the slope of the dose response curve. A more reliable method for quantitatively describing the changes in Al-h concentration with the dose would be using the simulated signals and calculating the area under the curve with double integration. This value is directly proportional to spin concentration and will not be affected by any contributions from other paramagnetic species. Despite these advantages, this method is very time consuming, demands more signal processing and is not always accessible. However, as mentioned previously, adjusting the simulated spectrum to the experimental one requires a reference point (or, to be precise, a second reference point, the first being the baseline), which in this case, was chosen as the centre of the peak around $g \approx 2.0603$ (see Figure 1). This provides the possibility of obtaining a reliable measurement of the amplitude simply by measuring the peak-to-baseline height of this peak of the experimental spectra, further referred to as "B". The values of B_{exp} and B_{sim} (Figure 2) will therefore always be, by definition, equal to each other for every example of coarse and fine spectra. This approach allows for a much more accurate representation of Al-h signal intensity for fine grains, and may also improve the measurement of coarse grains, particularly for younger samples.

To investigate the effect of this overestimation on the shape and slope of the dose response curve (DRC), two sets of DRCs were constructed for sample ROX 1.14 (coarse and fine grains), using amplitudes A_{exp} (DRC A) and B_{exp} (DRC B) (Figure 3a,b). A sum of two exponential functions was used to fit the datapoints.

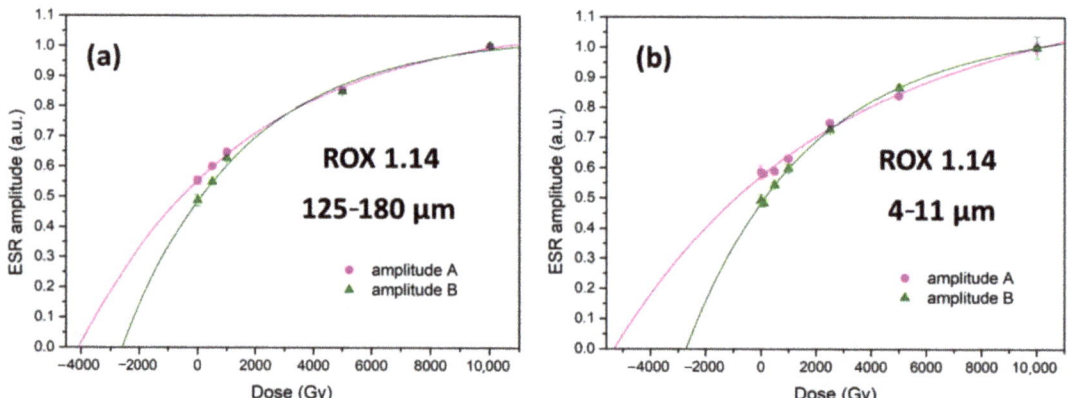

Figure 3. Dose response curves obtained for samples ROX 1.14 125–180 μm (**a**) and 4–11 μm (**b**) by measuring amplitudes A_{exp} and B_{exp}. Data normalised by the maximum value. Negative dose values indicate the dose accumulated in the material prior to laboratory irradiation.

As expected from the comparison between simulated and experimental spectra, the lower dose part of DRC A for coarse grains (Figure 3a) bends upwards compared to DRC B due to the contribution from peroxy signals, while at higher doses, curves A and B overlap. As a result, the equivalent dose obtained from DRC A is overestimated. The divergence between DRC A and B is more pronounced in the case of quartz fraction

4–11 μm (Figure 3b). Because of the peroxy contribution, DRC A has a much smaller slope, which leads to a considerable overestimation of the equivalent dose obtained from this curve compared to curve B. It should be kept in mind that, while both these curves seem to almost overlap at high doses, the values of amplitude A are still over 25% overestimated compared to amplitude B at 10,000 Gy. Since the exact nature of the dose-dependency of the signals observed in the case of fine grains is not known, the relationship between A and B values for doses above 10,000 Gy cannot be predicted at this point, and may further affect the shape and slope of DRC A.

3.1.2. Sample STY 1.10

The second example of comparison between Al-h measurements for coarse and fine grain spectra is sample STY 1.10 (Figure 4). The spectra of the 125–180 μm fraction demonstrate a similar situation to ROX 1.14—the shape of experimental spectra differs slightly from the simulation for smaller doses, and this difference becomes less significant as the radiation dose increases (Figure 4a). The value of A_{exp} amplitude overestimates A_{sim} by about 13% for the natural signal, and about 5% for 1000 Gy and 10,000 Gy. For fine grains represented by a 4–11 μm fraction (Figure 4b), the spectra for the natural sample and the sample irradiated with 1000 Gy show differences between simulation and experiment analogous to the ones observed for sample ROX 1.14. Due to the contribution of peroxy signals, A_{exp} overestimates A_{sim} by about 24% for the natural sample and about 21% for 1000 Gy irradiated sample. However, for higher doses, the situation becomes even more complex, as the overestimation of A_{exp} compared to A_{sim} increases again, to about 30% for 10,000 Gy. The explanation for this fact comes from analysing the peroxy signals observed at room temperature (Figure 4c,d). While the spectra of coarse grains (Figure 4c), as in the case of sample ROX 1.14 (Figure 2c), do not show significant changes as the dose increases, with only the Ge signal at $g \approx 1.996$ and E' signal at $g \approx 2.000$ being more prominent at high doses, the spectra of fine grains (Figure 4d) exhibit some additional signals, which increase their intensity with the laboratory dose. Most of them—the signal at $g \approx 2.002, 2.010$, the Ge signal ($g \approx 1.995$) and the E' signal ($g \approx 2.000$)—are also detected in sample ROX 1.14 (Figure 2d), but two other signals at $g \approx 1.991$ and 2.016 are not. In particular, the signal at $g \approx 2.016$ exhibits a considerable growth, and due to its position, which almost coincides with the top of the first peak of the central Al-h signal ($g = 2.018$) used for A_{exp} estimation, it strongly affects the outcome of this measurement for higher doses. As a result, A_{exp} amplitude obtained for fine grains provides unreliable measurements not only for lower doses, but also for higher doses, making it unsuitable for Al-h intensity determination. It is worth mentioning that, at first glance, the low-temperature spectrum of STY 1.10 irradiated with 10,000 Gy does not show clear signs of distortion around $g = 2.018$, as it still resembles the shape of the Al-h signal quite well, which can be very misleading, as it encourages attempts to measure A_{exp}. It is only through analysing the room temperature measurements that the dose-dependent nature of the signal at $g \approx 2.016$ can be revealed. In cases such as this, measuring A_{exp}, although technically possible, results in unreliable data, leading to a distorted shape in the dose-response curve. Amplitude B, however, remains unaffected by the contribution of other signals, and therefore provides a reliable representation of Al-h signal intensity changes.

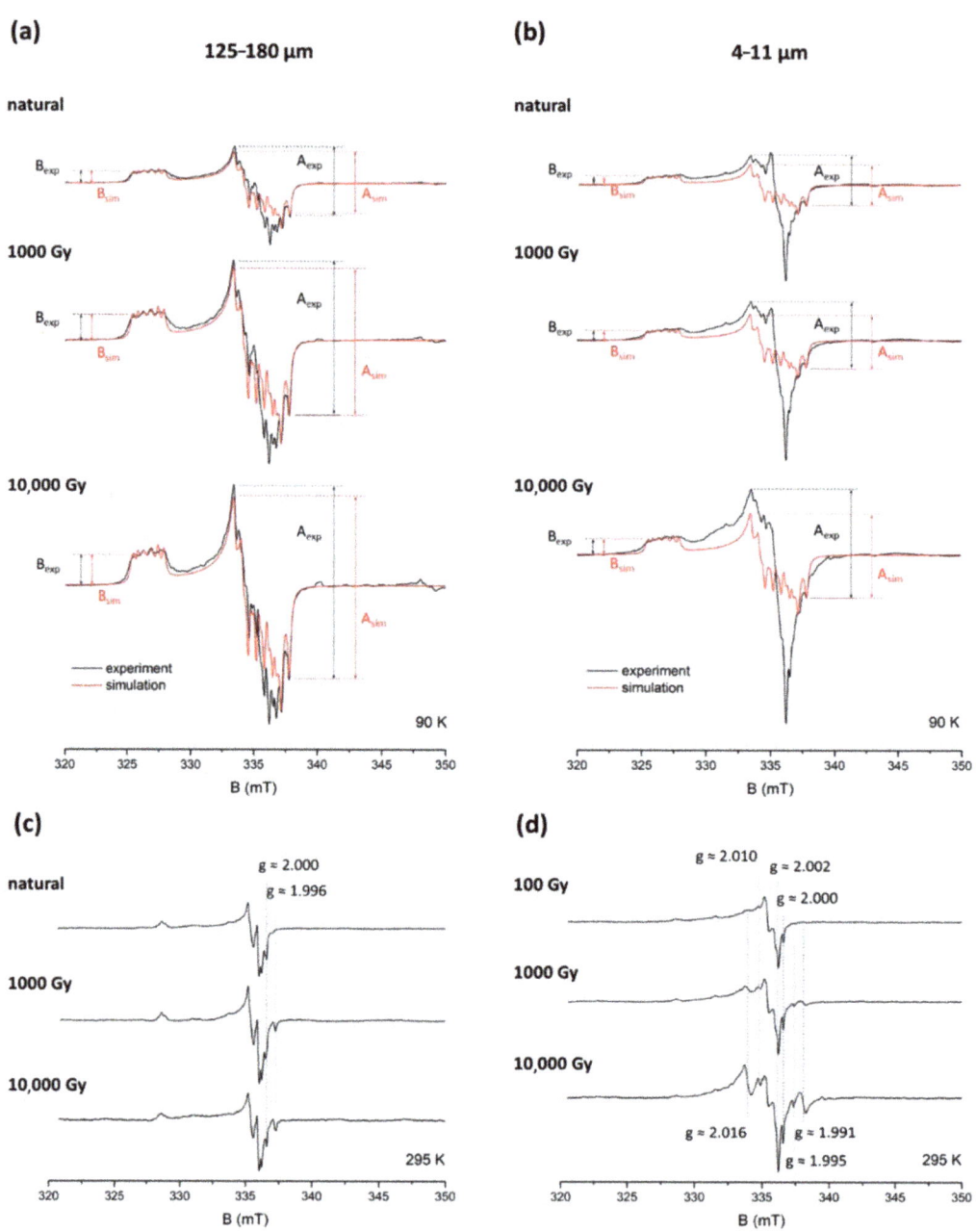

Figure 4. EPR spectra of the coarse (125–180 μm) and fine (4–11 μm) fraction of sample STY 1.10, natural and additionally irradiated with 1000 and 10,000 Gy, recorded at 90 K ((**a**,**b**), respectively) and at room temperature ((**c**,**d**), respectively), and simulated spectra of Al-h signal (in red). Amplitudes A_{exp}, A_{sim}, B_{exp} and B_{sim} are marked with arrows. Major dose-dependent signals observed at room temperature are marked with blue dashed lines. The 90 K spectra for coarse grains are multiplied by a factor of 5.9.

3.1.3. Sample 2 MV 80

The third example is based on measurements of sample 2 MV 80 (Figure 5). The spectra recorded at 90 K for coarse grains (Figure 5a), in this case represented by a 63–90 μm fraction, show a more significant distortion than in the case of samples ROX 1.14 and STY 1.14. This is most likely due to the smaller size of the coarse grains—63–90 μm instead of 125–180 μm—compared to the other two samples. As shown by Timar-Gabor [22], the intensity of peroxy signals decreases with increasing grain size. For the natural sample, A_{exp} overestimates A_{sim} value by as much as 82%, by 62% for 500 Gy and by 44% for 5000 Gy. A small part of this overestimation might be attributed to performing a baseline correction, as the original baselines displayed a steeper slope and more complex shape, but even then, the differences between A_{exp} and A_{sim} are still very considerable. Due to a limited availability of coarse material, fewer additional doses could be investigated; therefore, the overestimation present at 10,000 Gy could not be determined. The spectra recorded at room temperature resemble those obtained for samples ROX 1.14 and STY 1.10, with the same dose-dependent signals at $g \approx 2.000$ and $g \approx 1.996$, ascribed to the E' and Ge centre, respectively, being visible.

As with samples ROX 1.14 and STY 1.10, for fine grains (Figure 5b), the contribution of the peroxy signals in the central part of the spectrum is clearly visible, leading to an overestimation of A_{exp} by 50% compared to A_{sim} for the natural sample and 64% for the 1000 Gy irradiated sample. At the higher doses, as shown for 10,000 Gy, the spectrum becomes very distorted, to the point that the measurement of A_{exp} is basically impossible, as it is clearly too affected by the overlapping signals. Measurements performed at room temperature (Figure 5d) show the same dose-dependent signals, as in the case of the samples ROX.1.14 and STY.10—at $g \approx 2.010$, 2.002, and at $g \approx 1.999$ (E' centre) and 1.995 (Ge centre), as well as very strong dose-dependent signals at $g \approx 2.015$ and $g \approx 1.991$, also observed in sample STY 1.10, in addition to the non-dose-dependent signals, also visible in the coarse grains. In the case of samples like 2 MV 80, with very strong dose-dependent signals overlapping with the Al-h signal, the measurement of amplitude B is not only more reliable, but also appears to be the only viable option for obtaining Al-h amplitude without the use of simulations. It should be noted that the presence of dose-dependent signals will also cause problems when attempting to remove the peroxy signals by subtracting the spectra recorded after heating (as performed by Richer and Tsukamoto [36]), since the peroxy spectrum will look different for every dose, and the heating will likely affect the ratio of dose-dependent and non-dose-dependent peroxy signals. These arguments further support using amplitude B for Al-h intensity determination for both fine and coarse grains.

The comparison between the results obtained for the three presented examples by measuring the amplitude A and B shows the advantage of using amplitude B for Al-h intensity estimation. Contrary to amplitude A, it is not affected by the peroxy signals present in the centre of the analysed range, and it can therefore provide more accurate results, or, in fact, any results in cases where a spectrum is too distorted to allow for an estimation of amplitude A. Measurements of amplitude B were used in the second part of this study to compare the response of coarse and fine grains of quartz to laboratory irradiation.

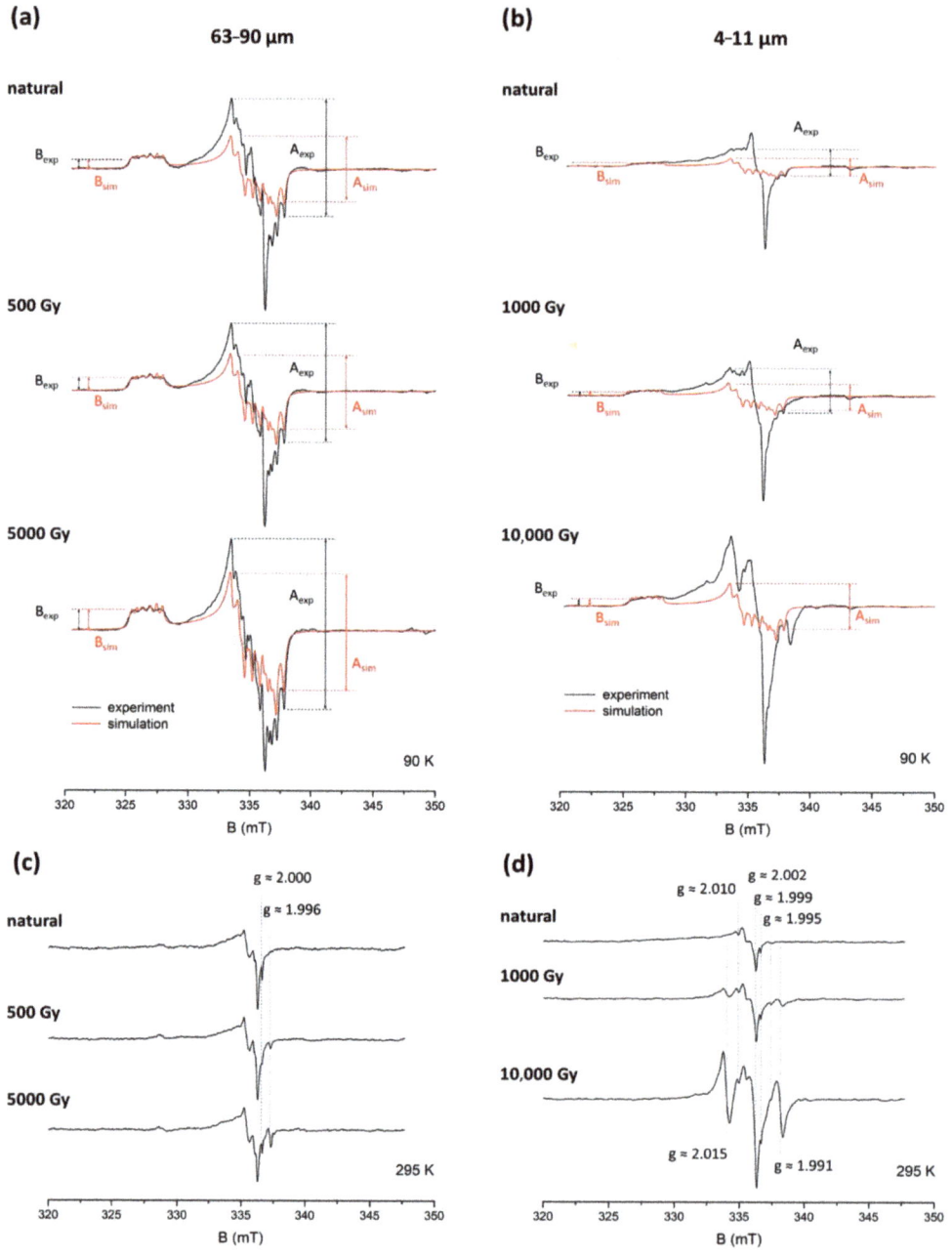

Figure 5. EPR spectra of the coarse (63–90 μm) and fine (4–11 μm) fraction of sample 2 MV 80, natural and additionally irradiated with 500 and 5000 Gy for coarse grains, and 1000 and 10,000 Gy for fine grains, recorded at 90 K ((**a**,**b**), respectively) and at room temperature ((**c**,**d**), respectively), and simulated spectra of Al-h signal (in red). Amplitudes A_{exp}, A_{sim}, B_{exp} and B_{sim} are marked with arrows. Major dose-dependent signals observed at room temperature are marked with blue dashed lines. The 90 K spectra for coarse grains are multiplied by a factor of 5.6.

3.2. Comparison of DRCs of Coarse and Fine Grains

Dose response curves obtained using amplitude B_{exp} were used for comparing the response of coarse and fine grains of quartz to laboratory irradiation (Figure 6). The behaviour of the Al-h signal was investigated up to 10,000 Gy on top of the natural dose for sample ROX 1.14, and up to 40,000 Gy for samples STY 1.10 and 2 MV 80. Due to the limited availability of the material, fewer aliquots were obtained for the coarse fraction than the fine fraction. No correction for the residual dose was applied.

A note of caution regarding the fitting is necessary before proceeding to describe these results. A sum of two saturating exponential functions was used to fit the data. This choice was dictated by the results obtained in a recent study by Benzid and Timar-Gabor [43], where a phenomenological model of Al-h formation upon room temperature irradiation was proposed. In this model, the Al-h centre is considered to be formed upon laboratory irradiation by two processes: (i) directly by transforming $[AlO_4/M^+]^0$ into Al-h, and (ii) indirectly by transforming $[AlO_4/M^+]^0$ into $[AlO_4/H^+]^0$, and then $[AlO_4/H^+]^0$ into Al-h. By assuming that the dissociation rates of these centres are proportional to their concentrations, the model shows that the increase in the Al-h EPR signal with increasing dose can be well described by a sum of two exponential functions. Benzid and Timar-Gabor, however, acknowledge the dangers of interpreting the parameters derived through fitting with multiple exponentials, stating that, for quantitative assumptions using the derived parameters to be made, the DRC needs to be raised until it reaches full saturation; otherwise, the parameters depend on the maximum given dose, as was shown previously by Timar-Gabor et al. [21] for DRCs obtained for OSL signals fitted with a sum of two saturating exponentials. As is clear from Figure 6, this is not the case for DRCs constructed in the current study; therefore, we refrain from deriving any conclusions based on parameters obtained from the fittings. As such, the fitted curves presented in Figure 6 should be regarded primarily as a visual aid in comparing the response to laboratory irradiation. Additionally, due to a smaller number of datapoints for coarse-grained samples STY 1.10 and 2 MV 80 and their noticeable scatter, a fitting was not performed.

Proceeding to the comparison of fine and coarse quartz DRCs, it is immediately apparent from Figure 6a that both the 4–11 µm and 125–180 µm fractions of the ROX 1.14 sample show almost identical DRC shape. While the number of datapoints is limited, it can be assumed that the effect of increasing the laboratory dose on the Al-h signal, even if not identical, is remarkably similar in fine and coarse grains. As mentioned before, the data obtained for the coarse fraction of samples STY 1.10 (Figure 6b) 2 MV 80 (Figure 6c) did not allow for a satisfactory fitting, as the shape of the fitted curves would be largely affected by an arbitrary choice of parameters. Instead, the datapoints were overlaid on the DRCs obtained for fine grains. Some differences can be observed for the sample STY 1.10 (Figure 6b), namely, the datapoints obtained for coarse (125–180 µm) grains seem to indicate a faster saturation of the DRC. However, the shape of the curves is likely affected by a noticeable scatter of the datapoints and the absence of data for coarse grains above 20,000 Gy, so the divergence observed in Figure 6b may very well be exaggerated. In the case of sample 2 MV 570 (Figure 6c), as far as the doses up to 5000 Gy are concerned, the data for coarse (63–90 um) and fine grains is in very good agreement, suggesting that, in this range, there are no significant differences in the behaviour of the Al-h signal in coarse and fine grains for this sample. Despite the aforementioned issues with the fitting, simply by visually following the datapoints, it can be observed that, in all three cases, the intercept of the DRCs with the x axis seem to be the same for both fractions.

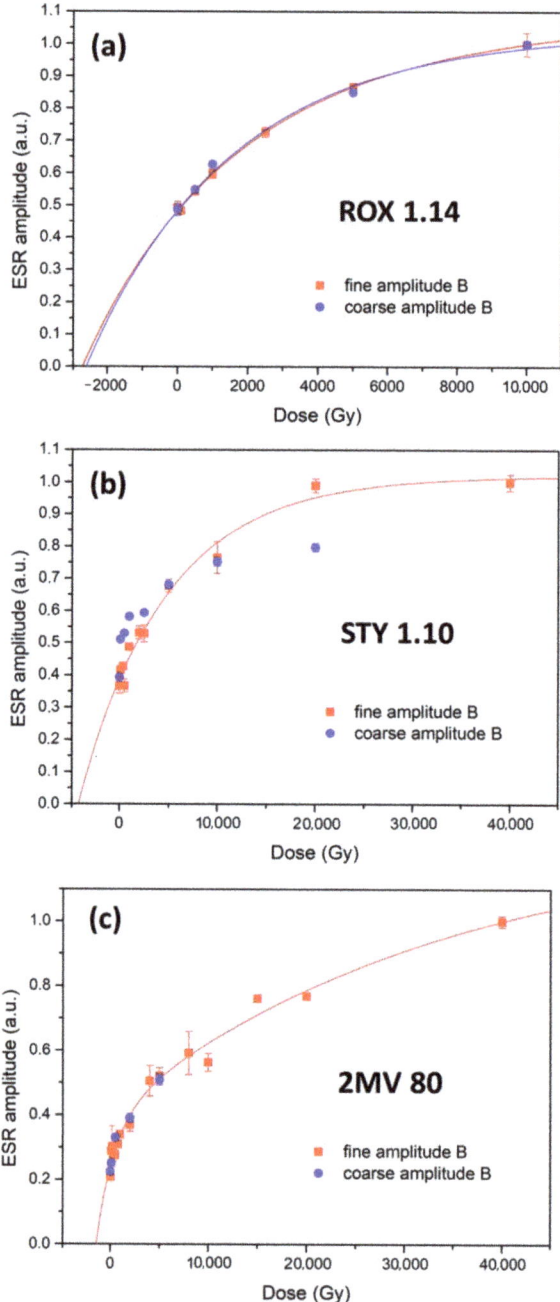

Figure 6. Dose response curves for the amplitude B_{exp} obtained for fine (4–11 µm) and coarse (125–180 µm) grains of samples ROX 1.14 (**a**), STY 1.10 (**b**), and fine (4–11 µm) and coarse (63–90 µm) grains of sample 2 MV 80 (**c**). Data were normalised to the maximum datapoint (**a**) or maximum datapoint for fine grains and overlaid on the curve for coarse grains (**b**,**c**). Negative dose values indicate the dose accumulated in the material prior to laboratory irradiation.

It can therefore be stated that no significant divergence in the behaviour of the Al-h signal with the increasing radiation dose in the investigated range can be observed between the fine (4–11 μm) and coarse (63–90 μm and 125–180 μm) grains of quartz studied in this work. A logical conclusion is to assume that Al-h centre formation in fine and coarse grains due to artificial radiation is governed by the same processes. To our knowledge, the influence of grain size on the formation of the Al-h centre has not been discussed in the literature. The phenomenological model of Al-h formation upon laboratory irradiation at room temperature proposed by Benzid and Timar-Gabor [43] does not suggest that this process would be significantly different for coarse and fine grains. As Al-h centres are extrinsic, impurity-related defects, it is to be expected that they would have a relatively homogeneous distribution in the volume of a sedimentary quartz grain. Indeed, Timar-Gabor [22] report that no significant effect could be observed when measuring Al-h signals as a function of etching time. More experimental and theoretical studies are certainly needed to further examine the mechanism of Al-h centre formation; however, our results show that, in the first approximation, the response of both fine- and coarse-grained quartz to artificial irradiation is remarkably similar.

Should coarse and fine grains of quartz therefore be provided with the same equivalent dose when dated using the Al-h centre? The answer to this question requires a separate consideration and cannot be answered at this point. It is generally accepted that sunlight exposure does not completely bleach the EPR intensity of Al-h, and the signal is reset only to a non-bleachable residual level (e.g., [6,7,44–48]). Our study focuses on unbleached samples, which have a residual signal composed of bleachable and unbleachable components. These components can be of different magnitudes for coarse and fine grains and should be determined separately for every fraction, which is beyond the scope of this work. To our knowledge, the only study showing the effects of grain size on the results of the EPR dating of quartz was conducted by Liu et al. [49] for the Ti-Li centres of fluvial and lacustrine sediments. They assumed complete bleaching and reported that, for grain sizes above 100 μm, the equivalent dose decreased with the increase in grain size. However, for the smallest fraction (50–100 μm, which, in our study, would still be considered coarse), the equivalent dose was smaller than for the larger fraction. They also showed that the beta irradiation dose rate of grains with different sizes accounts for only about 6% of the total deviation of dating results, making if far less significant than the effect of grain size on EPR sensitivity. No similar studies have been conducted on the Al-h centre of sedimentary quartz. It should be mentioned that the effect that the size of grains has on the obtained equivalent dose was investigated for E' and the Al-h centre in quartz from fault gouge (e.g., [50,51]), but due to the different mechanism involved in resetting the signal (mechanical deformation and high temperature), these results cannot be of use for other types of environments. The effects of natural irradiation and light exposure on fine- and coarse-grained quartz should certainly be investigated in order to reach conclusions regarding the equivalent dose estimation.

4. Conclusions

We examined the Al-h and peroxy EPR spectra of fine (4–11 μm) and coarse (63–90, 125–180 μm) sedimentary quartz separates extracted from three well-characterised samples collected from thoroughly investigated sites (Roxolany, Stayky and Mircea Vodă). Based on the data presented in this work, as well as in the study conducted by Timar-Gabor [22], it is clear that Al-h measurements of fine grains are affected by the presence of peroxy signals to a much greater extent than coarse grains. However, the degree to which this affects the standard amplitude measurement following the approach of Toyoda and Falguères [27] seems to be sample-dependent. It ranges from causing an overestimation, which is much stronger for smaller doses (sample ROX 1.14), to a complete distortion of the spectra at high doses (sample 2 MV 80) due to the presence of dose-dependent peroxy signals in fine grains. For a proper understanding of the observed differences, a much larger set of samples would certainly have to be analysed, which is not an easy task, since it requires a large amount of material from different sites divided into fractions.

The new approach to measuring Al-h signal amplitude proposed in this study, focusing on the peak-to-baseline amplitude of the part of the signal at $g \approx 2.0603$, has the potential to provide more accurate results. This region of the spectrum is not affected by strong peroxy signals overlapping the central part of the Al-h signal and causing the overestimation. While using the strongest absorption line, as in the standard approach, increases the signal-to-noise ratio, which leads to greater precision in the dating result, our study shows that this precision comes at the cost of accuracy. In other words, while the errors associated with the standard approach may be smaller, the dates themselves may not reflect the true age of the material. We believe that more accurate results, even if less precise, are of much greater value to the dating community and the researchers using the reported values in their studies. It should be mentioned that, while very useful for samples with strong Al-h signals, as the ones investigated here, the new approach might not be applicable to very weak signals. This part of the Al-h spectrum is considerably less intense than the central signal typically used for measuring amplitude, and in the case of some samples, it may simply be undetectable. Additionally, more studies are needed on the individual signals composing the peroxy spectra in order to rule out the possibility of some weaker lower-field peaks being present around $g \approx 2.0603$, which could affect the amplitude measurement following this new approach.

We compared the response of the Al-h signal to laboratory irradiation displayed by the fine- and coarse-grained fractions, which has not been previously shown in the literature. The shapes of dose response curves constructed for coarse and fine grains using the new approach show a considerable similarity, which suggests that Al-h centre formation in fine and coarse grains upon artificial radiation follows the same pattern. These observations have significant implications for the dating community and will hopefully inspire more research, experimental and theoretical, allowing for a thorough comparison of dating results obtained for different fractions of sedimentary quartz, which in turn will deepen our understanding of the underlying processes and increase the accuracy of EPR dating.

It should be stressed that the behaviour of the Al-h signal in coarse and fine grains upon laboratory irradiation might differ from behaviour observed in nature. Depending on the grain size, the amount of alpha and beta radiation penetrating the grain will be different, influencing the formation of defects. Understanding the processes induced in fine and coarse grains by gamma radiation in a controlled laboratory environment is the first step towards the development of a comprehensive model. The effect of grain size on the formation and bleachability of Al-h centres under natural conditions needs to be thoroughly studied before any conclusions are drawn regarding the overall result of EPR dating using different fractions. We hope that our work will stimulate such studies.

Author Contributions: Conceptualization, Z.K. and A.T.-G. Investigation, Z.K. and A.T.-G. Writing—Original draft preparation, Z.K. Writing—Review and editing, A.T.-G. Visualization, Z.K. Supervision, A.T.-G. Project administration, A.T.-G. Funding acquisition, A.T.-G. All authors have read and agreed to the published version of the manuscript.

Funding: This study received funding from the European Research Council (ERC) under the European Union's Horizon 2020 research and innovation programme, ERC-2015-STG (grant agreement No [678106]). A.T.-G. acknowledges the financial support of the research project EEA-RO-NO-2018-0126.

Institutional Review Board Statement: Not applicable.

Informed Consent Statement: Not applicable.

Data Availability Statement: The data analysed in this study are available from the corresponding authors upon reasonable request.

Conflicts of Interest: The authors declare no conflict of interest.

References

1. Preusser, F.; Chithambo, M.L.; Götte, T.; Martini, M.; Ramseyer, K.; Sendezera, E.J.; Susino, G.J.; Wintle, A.G. Quartz as a Natural Luminescence Dosimeter. *Earth-Sci. Rev.* **2009**, *97*, 184–214. [CrossRef]
2. Götze, J. Chemistry, Textures and Physical Properties of Quartz—Geological Interpretation and Technical Application. *Mineral. Mag.* **2009**, *73*, 645–671. [CrossRef]
3. Malik, D.M.; Kohnke, E.E.; Sibley, W.A. Low-temperature Thermally Stimulated Luminescence of High Quality Quartz. *J. Appl. Phys.* **1981**, *52*, 3600–3605. [CrossRef]
4. Toyoda, S. Paramagnetic Lattice Defects in Quartz for Applications to ESR Dating. *Quat. Geochronol.* **2015**, *30*, 498–505. [CrossRef]
5. Ikeya, M. *New Applications of Electron Spin Resonance: Dating, Dosimetry and Microscopy*; World Scientific: Singapore; Hackensack, NJ, USA; London, UK; Hong Kong, China, 1993.
6. Duval, M.; Arnold, L.J.; Guilarte, V.; Demuro, M.; Santonja, M.; Pérez-González, A. Electron Spin Resonance Dating of Optically Bleached Quartz Grains from the Middle Palaeolithic Site of Cuesta de La Bajada (Spain) Using the Multiple Centres Approach. *Quat. Geochronol.* **2017**, *37*, 82–96. [CrossRef]
7. Tsukamoto, S.; Long, H.; Richter, M.; Li, Y.; King, G.E.; He, Z.; Yang, L.; Zhang, J.; Lambert, R. Quartz Natural and Laboratory ESR Dose Response Curves: A First Attempt from Chinese Loess. *Radiat. Meas.* **2018**, *120*, 137–142. [CrossRef]
8. Timar-Gabor, A.; Vandenberghe, D.A.G.; Vasiliniuc, S.; Panaoitu, C.E.; Panaiotu, C.G.; Dimofte, D.; Cosma, C. Optical Dating of Romanian Loess: A Comparison between Silt-Sized and Sand-Sized Quartz. *Quat. Int.* **2011**, *240*, 62–70. [CrossRef]
9. Tissoux, H.; Toyoda, S.; Falguères, C.; Voinchet, P.; Takada, M.; Bahain, J.-J.; Despriée, J. ESR Dating of Sedimentary Quartz from Two Pleistocene Deposits Using Al and Ti-Centers. *Geochronometria* **2008**, *30*, 23–31. [CrossRef]
10. Weil, J.A. A Review of Electron Spin Spectroscopy and Its Application to the Study of Paramagnetic Defects in Crystalline Quartz. *Phys. Chem. Miner.* **1984**, *10*, 149–165. [CrossRef]
11. Duval, M.; Guilarte, V. ESR Dosimetry of Optically Bleached Quartz Grains Extracted from Plio-Quaternary Sediment: Evaluating Some Key Aspects of the ESR Signals Associated to the Ti-Centers. *Radiat. Meas.* **2015**, *78*, 28–41. [CrossRef]
12. Laurent, M.; Falguères, C.; Bahain, J.; Rousseau, L.; Van Vliet Lanoé, B. ESR Dating of Quartz Extracted from Quaternary and Neogene Sedimentsmethod, Potential and Actual Limits. *Quat. Sci. Rev.* **1998**, *17*, 1057–1062. [CrossRef]
13. Parés, J.M.; Álvarez, C.; Sier, M.; Moreno, D.; Duval, M.; Woodhead, J.D.; Ortega, A.I.; Campaña, I.; Rosell, J.; Bermúdez de Castro, J.M.; et al. Chronology of the Cave Interior Sediments at Gran Dolina Archaeological Site, Atapuerca (Spain). *Quat. Sci. Rev.* **2018**, *186*, 1–16. [CrossRef]
14. Rink, W.J.; Bartoll, J.; Schwarcz, H.P.; Shane, P.; Bar-Yosef, O. Testing the Reliability of ESR Dating of Optically Exposed Buried Quartz Sediments. *Radiat. Meas.* **2007**, *42*, 1618–1626. [CrossRef]
15. Tissoux, H.; Voinchet, P.; Lacquement, F.; Prognon, F.; Moreno, D.; Falguères, C.; Bahain, J.-J.; Toyoda, S. Investigation on Non-Optically Bleachable Components of ESR Aluminium Signal in Quartz. *Radiat. Meas.* **2012**, *47*, 894–899. [CrossRef]
16. Voinchet, P.; Bahain, J.J.; Falguères, C.; Laurent, M.; Dolo, J.M.; Despriée, J.; Gageonnet, R.; Chaussé, C. ESR Dating of Quartz Extracted from Quaternary Sediments Application to Fluvial Terraces System of Northern France [Datation Par Résonance Paramagnétique Électronique (RPE) de Quartz Fluviatiles Quaternaires: Application Aux Systèmes de Terrasses Du Nord]. *Quaternaire* **2004**, *15*, 135–141. [CrossRef]
17. Voinchet, P.; Yin, G.; Falguères, C.; Liu, C.; Han, F.; Sun, X.; Bahain, J.J. Dating of the Stepped Quaternary Fluvial Terrace System of the Yellow River by Electron Spin Resonance (ESR). *Quat. Geochronol.* **2019**, *49*, 278–282. [CrossRef]
18. Moreno, D.; Duval, M.; Rubio-Jara, S.; Panera, J.; Bahain, J.J.; Shao, Q.; Pérez-González, A.; Falguères, C. ESR Dating of Middle Pleistocene Archaeo-Paleontological Sites from the Manzanares and Jarama River Valleys (Madrid Basin, Spain). *Quat. Int.* **2019**, *520*, 23–38. [CrossRef]
19. Moreno, D.; Gutiérrez, F.; del Val, M.; Carbonel, D.; Jiménez, F.; Jesús Alonso, M.; Martínez-Pillado, V.; Guzmán, O.; López, G.I.; Martínez, D. A Multi-Method Dating Approach to Reassess the Geochronology of Faulted Quaternary Deposits in the Central Sector of the Iberian Chain (NE Spain). *Quat. Geochronol.* **2021**, *65*, 101185. [CrossRef]
20. Beerten, K.; Lomax, J.; Clémer, K.; Stesmans, A.; Radtke, U. On the Use of Ti Centres for Estimating Burial Ages of Pleistocene Sedimentary Quartz: Multiple-Grain Data from Australia. *Quat. Geochronol.* **2006**, *1*, 151–158. [CrossRef]
21. Timar-Gabor, A.; Buylaert, J.-P.; Guralnik, B.; Trandafir-Antohi, O.; Constantin, D.; Anechitei-Deacu, V.; Jain, M.; Murray, A.S.; Porat, N.; Hao, Q.; et al. On the Importance of Grain Size in Luminescence Dating Using Quartz. *Radiat. Meas.* **2017**, *106*, 464–471. [CrossRef]
22. Timar-Gabor, A. Electron Spin Resonance Characterisation of Sedimentary Quartz of Different Grain Sizes. *Radiat. Meas.* **2018**, *120*, 59–65. [CrossRef]
23. Yokoyama, Y.; Falgueres, C.; Quaegebeur, J.P. ESR Dating of Quartz from Quaternary Sediments: First Attempt. *Nucl. Tracks Radiat. Meas.* **1985**, *10*, 921–928. [CrossRef]
24. Toyoda, S.; Ikeya, M. ESR Dating of Quartz and Plagioclase from Volcanic Ashes Using E'1, A1 and Ti Centers. *Int. J. Radiat. Appl. Instrumentation. Part D. Nucl. Tracks Radiat. Meas.* **1991**, *18*, 179–184. [CrossRef]
25. Imai, N.; Shimokawa, K. ESR Dating of Quaternary Tephra from Mt. Osore-Zan Using Al and Ti Centres in Quartz. *Quat. Sci. Rev.* **1988**, *7*, 523–527. [CrossRef]
26. Lin, M.; Yin, G.; Ding, Y.; Cui, Y.; Chen, K.; Wu, C.; Xu, L. Reliability Study on ESR Dating of the Aluminum Center in Quartz. *Radiat. Meas.* **2006**, *41*, 1045–1049. [CrossRef]

27. Toyoda, S.; Falguères, C. The Method to Represent the ESR Signal Intensity of the Aluminum Hole Center in Quartz for the Purpose of Dating. *Adv. ESR Appl.* **2003**, *20*, 7–10.
28. Friebele, E.J.; Griscom, D.L.; Stapelbroek, M.; Weeks, R.A. Fundamental Defect Centers in Glass: The Peroxy Radical in Irradiated, High-Purity, Fused Silica. *Phys. Rev. Lett.* **1979**, *42*, 1346–1349. [CrossRef]
29. Botis, S.; Nokhrin, S.M.; Pan, Y.; Xu, Y.; Bonli, T.; Sopuck, V. Natural Radiation-Induced Damage in Quartz. I. Correlations between Cathodoluminence Colors and Paramagnetic Defects. *Can. Mineral.* **2005**, *43*, 1565–1580. [CrossRef]
30. Botis, S.M.; Pan, Y.; Nokhrin, S.; Nilges, M.J. Natural Radiation-Induced Damage in Quartz. III. A New Ozonide Radical in Drusy Quartz from the Athabasca Basin, Saskatchewan. *Can. Miner.* **2008**, *46*, 125–138. [CrossRef]
31. Nilges, M.J.; Pan, Y.; Mashkovtsev, R.I. Radiation-Damage-Induced Defects in Quartz. I. Single-Crystal W-Band EPR Study of Hole Centers in an Electron-Irradiated Quartz. *Phys. Chem. Miner.* **2008**, *35*, 103–115. [CrossRef]
32. Nilges, M.J.; Pan, Y.; Mashkovtsev, R.I. Radiation-Induced Defects in Quartz. III. Single-Crystal EPR, ENDOR and ESEEM Study of a Peroxy Radical. *Phys. Chem. Miner.* **2009**, *36*, 61–73. [CrossRef]
33. Pan, Y.; Nilges, M.J.; Mashkovtsev, R.I. Radiation-Induced Defects in Quartz. II. Single-Crystal W-Band EPR Study of a Natural Citrine Quartz. *Phys. Chem. Miner.* **2008**, *35*, 387–397. [CrossRef]
34. Pan, Y.; Nilges, M.J.; Mashkovtsev, R.I. Radiation-Induced Defects in Quartz: A Multifrequency EPR Study and DFT Modelling of New Peroxy Radicals. *Mineral. Mag.* **2009**, *73*, 519–535. [CrossRef]
35. Skuja, L.; Ollier, N.; Kajihara, K. Luminescence of Non-Bridging Oxygen Hole Centers as a Marker of Particle Irradiation of α-Quartz. *Radiat. Meas.* **2020**, *135*, 106373. [CrossRef]
36. Richter, M.; Tsukamoto, S. Investigation of Quartz Electron Spin Resonance Residual Signals in the Last Glacial and Early Holocene Fluvial Deposits from the Lower Rhine. *Geochronology* **2022**, *4*, 55–63. [CrossRef]
37. Anechitei-Deacu, V.; Timar-Gabor, A.; Thomsen, K.J.; Buylaert, J.-P.; Jain, M.; Bailey, M.; Murray, A.S. Single and Multi-Grain OSL Investigations in the High Dose Range Using Coarse Quartz. *Radiat. Meas.* **2018**, *120*, 124–130. [CrossRef]
38. Veres, D.; Tecsa, V.; Gerasimenko, N.; Zeeden, C.; Hambach, U.; Timar-Gabor, A. Short-Term Soil Formation Events in Last Glacial East European Loess, Evidence from Multi-Method Luminescence Dating. *Quat. Sci. Rev.* **2018**, *200*, 34–51. [CrossRef]
39. Groza-Săcaciu, Ș.-M.; Panaiotu, C.; Timar-Gabor, A. Single Aliquot Regeneration (SAR) Optically Stimulated Luminescence Dating Protocols Using Different Grain-Sizes of Quartz: Revisiting the Chronology of Mircea Vodă Loess-Paleosol Master Section (Romania). *Methods Protoc.* **2020**, *3*, 19. [CrossRef]
40. Timar, A.; Vandenberghe, D.; Panaiotu, E.C.; Panaiotu, C.G.; Necula, C.; Cosma, C.; van den Haute, P. Optical Dating of Romanian Loess Using Fine-Grained Quartz. *Quat. Geochronol.* **2010**, *5*, 143–148. [CrossRef]
41. Stoll, S.; Schweiger, A. EasySpin, a Comprehensive Software Package for Spectral Simulation and Analysis in EPR. *J. Magn. Reson.* **2006**, *178*, 42–55. [CrossRef]
42. Nuttall, R.H.D.; Weil, J.A. The Magnetic Properties of the Oxygen–Hole Aluminum Centers in Crystalline SiO_2. I. $[AlO_4]^0$. *Can. J. Phys.* **1981**, *59*, 1696–1708. [CrossRef]
43. Benzid, K.; Timar-Gabor, A. Phenomenological Model of Aluminium-Hole ($[AlO_4/H+]0$) Defect Formation in Sedimentary Quartz upon Room Temperature Irradiation: Electron Spin Resonance (ESR) Study. *Radiat. Meas.* **2020**, *130*, 106187. [CrossRef]
44. Timar-Gabor, A.; Chruścińska, A.; Benzid, K.; Fitzsimmons, K.E.; Begy, R.; Bailey, M. Bleaching Studies on Al-Hole ($[AlO_4/h]_0$) Electron Spin Resonance (ESR) Signal in Sedimentary Quartz. *Radiat. Meas.* **2020**, *130*, 106221. [CrossRef]
45. Tissoux, H.; Falguères, C.; Voinchet, P.; Toyoda, S.; Bahain, J.J.; Despriée, J. Potential Use of Ti-Center in ESR Dating of Fluvial Sediment. *Quat. Geochronol.* **2007**, *2*, 367–372. [CrossRef]
46. Toyoda, S.; Voinchet, P.; Falguères, C.; Dolo, J.M.; Laurent, M. Bleaching of ESR Signals by the Sunlight: A Laboratory Experiment for Establishing the ESR Dating of Sediments. *Appl. Radiat. Isot.* **2000**, *52*, 1357–1362. [CrossRef]
47. Tsukamoto, S.; Porat, N.; Ankjærgaard, C. Dose Recovery and Residual Dose of Quartz ESR Signals Using Modern Sediments: Implications for Single Aliquot ESR Dating. *Radiat. Meas.* **2017**, *106*, 472–476. [CrossRef]
48. Voinchet, P.; Falguères, C.; Laurent, M.; Toyoda, S.; Bahain, J.J.; Dolo, J.M. Artificial Optical Bleaching of the Aluminium Center in Quartz Implications to ESR Dating of Sediments. *Quat. Sci. Rev.* **2003**, *22*, 1335–1338. [CrossRef]
49. Liu, C.-R.; Yin, G.-M.; Han, F. Effects of Grain Size on Quartz ESR Dating of Ti–Li Center in Fluvial and Lacustrine Sediments. *Quat. Geochronol.* **2015**, *30*, 513–518. [CrossRef]
50. Buhay, W.M.; Schwarcz, H.P.; Grün, R. ESR Dating of Fault Gouge: The Effect of Grain Size. *Quat. Sci. Rev.* **1988**, *7*, 515–522. [CrossRef]
51. Lee, H.-K.; Yang, J.-S. ESR Dating of the Wangsan Fault, South Korea. *Quat. Sci. Rev.* **2003**, *22*, 1339–1343. [CrossRef]

Article

Electron Spin Resonance Dosimetry Studies of Irradiated Sulfite Salts

Amanda Burg Rech [1], Angela Kinoshita [2,*], Paulo Marcos Donate [3], Otaciro Rangel Nascimento [4] and Oswaldo Baffa [1]

1   Departamento de Física, Faculdade de Filosofia Ciências e Letras de Ribeirão Preto, Universidade de São Paulo, Ribeirão Preto 14040-900, SP, Brazil
2   Pró-Reitoria de Pesquisa e Pós-Graduação, Universidade do Oeste Paulista, Presidente Prudente 19067-175, SP, Brazil
3   Departamento de Química, Faculdade de Filosofia Ciências e Letras de Ribeirão Preto, Universidade de São Paulo, Ribeirão Preto 14040-900, SP, Brazil
4   Departamento de Física Interdisciplinar, Instituto de Física de São Carlos, Universidade de São Paulo, São Carlos 13566-590, SP, Brazil
*   Correspondence: angela@unoeste.br; Tel.: +55-18-3229-2079

Abstract: The study of new materials for radiation dosimetry is important to improve the present state of the art and to help in cases of accidents for retrospective dosimetry. Sulfites are compounds that contain a sulfur ion, widely used in the food industry. Due to the significant application of these compounds, sulfites are interesting candidates for accidental dosimetry, as fortuitous radiation detectors. The presence of the SO_3^- anion enables its detection by electron spin resonance (ESR) spectroscopy. The Dose–Response behavior, signal stability and other spectral features were investigated for sodium sulfite, sodium bisulfite, sodium metabisulfite and potassium metabisulfite, all in crystalline forms. The ESR spectrum of salts presented stability and proportional response with dose, presenting potential for dosimetry applications.

Keywords: radiation dosimetry; sulfite; radiation accidents; retrospective dosimetry

Citation: Rech, A.B.; Kinoshita, A.; Donate, P.M.; Nascimento, O.R.; Baffa, O. Electron Spin Resonance Dosimetry Studies of Irradiated Sulfite Salts. *Molecules* **2022**, *27*, 7047. https://doi.org/10.3390/molecules27207047

Academic Editors: Yordanka Karakirova and Nicola D. Yordanov

Received: 1 September 2022
Accepted: 17 October 2022
Published: 19 October 2022

Publisher's Note: MDPI stays neutral with regard to jurisdictional claims in published maps and institutional affiliations.

Copyright: © 2022 by the authors. Licensee MDPI, Basel, Switzerland. This article is an open access article distributed under the terms and conditions of the Creative Commons Attribution (CC BY) license (https://creativecommons.org/licenses/by/4.0/).

1. Introduction

Sulfites are chemical compounds widely used in the food industry as a preservative and additive. They are sources of SO_2, which is an antimicrobial agent known since antiquity [1].

Parts of sulfites are the bisulfites (HSO_3^-) and metabisulfites ($S_2O_5^{2-}$), in which the anions SO_3^{2-}, HSO^{3-} and $S_2O_5^{2-}$ [2] are present and detectable by electron spin resonance (ESR) spectroscopy, what makes them potential materials to compose dosimeters. Because they are easily found in different applications and are in widespread use, they are a candidate for radiation dosimetry in cases of accidents, in pure form at the location. They can also be employed as a component of a dosimeter or blended with a binder, as is done with alanine, for some specific applications.

The SO_2^- and SO_3^- anions are commonly cited in ESR papers about speleothems, calcified tissues of shells, snails and corals, because they are generated by ionizing radiation. They are present in the structure of calcites and aragonites as impurities. In these materials, they are characterized by having stability, allowing their use to differ between irradiated from non-irradiated material [3] and other applications such as dosimetry and dating [4–8].

Other inorganic compounds containing the $SO_3 \cdot^-$ radical ion have already been studied as ESR dosimeters. Bogushevich and Ugolev [9] have discussed some aspects of inorganic ESR dosimeters for medical radiotherapy and have shown that alkaline earth dithionates ($S_2O_6^{2-}$) have great potential. Irradiated dithionates exhibit a narrow line stable at room temperature attributed to $SO_3 \cdot^-$ radical anions [10,11]. The oxidation of dithionates ($S_2O_6^{2-}$) could result in sulfites (SO_3^{2-}) and dithionites ($S_2O_4^{2-}$) [12], also detectable by ESR spectroscopy.

In this work, the spectroscopic characterization of the irradiated sulfites, the behavior of the signal as a function of microwave power, the stability of the signal and the properties of the ESR spectrum of irradiated sulfites as a function of dose of radiation are presented, showing that these compounds have the potential for dosimetric applications. Thus, the goal of this paper is twofold; first, to offer a characterization of the radicals created by ionizing radiation produced by X-ray sources in these compounds by simulating the ESR spectra, and to present the possibility of using these materials as fortuitous dosimeters. To achieve this possible emergency application, the dose response was studied for doses below 20 Gy. In this scenario, materials present at the local site of a radioactive accident that produces stable radicals can be used as a dosimeter giving valuable information to manage the situation.

2. Results and Discussion

2.1. ESR Spectra Characterization

Figure 1 shows the ESR spectra of samples before irradiation. The background signal may be due to intrinsic defects present in the samples, which were commercial salts and used as received without any further treatment to simulate a practical situation, and can be used as a fortuitous dosimeter. The spectra of sulfites are shown with the Mn^{2+} marker that was used for determining the g-factor. The radicals present in sodium sulfite, sodium bisulfite and sodium metabisulfite are characterized by a relatively simple ESR spectrum, with the g-factor around g = 2.0085 and a width of 0.4 mT. Potassium metabisulfite presents an asymmetric line leading to the hypothesis of the presence of two or more radicals or crystal orientation in relation to the magnetic field.

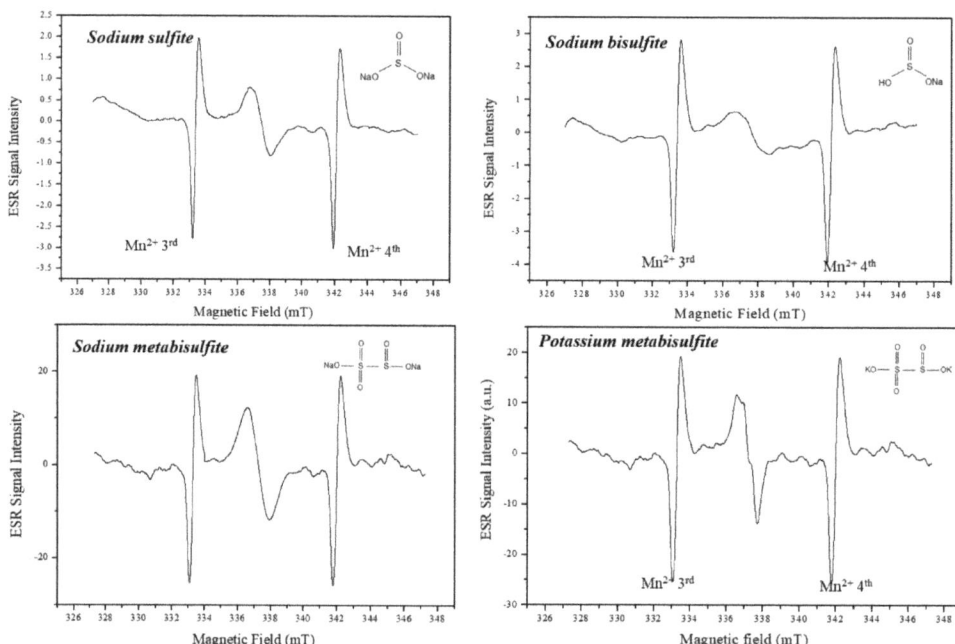

Figure 1. ESR signal of non-irradiated sulfites and their chemical structure. Spectra were recorded with Mn marker.

As already mentioned, potassium metabisulfite appears to have a composite line structure, different from the other sulfites studied. Thus, Figure 2 shows the simulation

(EasySpin [13]) of the sulfites studied after irradiation with a dose of 500 Gy for comparison. All radicals identified with 500 Gy fitted in the lower dose spectra. Table 1 lists the parameters found in the spectra simulation.

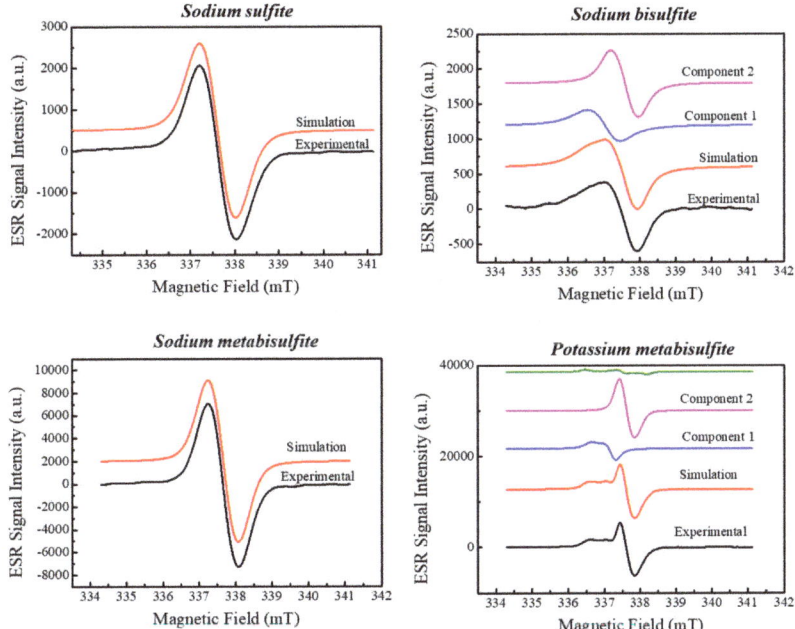

Figure 2. Simulation of ESR spectra of sulfites irradiated with 500 Gy.

Table 1. Parameters of Spin Hamiltonian of sulfites, g-factor and linewidth (LW).

Compound	g-Factor	LW (mT)
Sodium sulfite	2.0085 2.0062 2.0062	0.6266 0.2427
Sodium bisulfite	2.0133 2.0089 2.0098 2.0047 2.0071 2.0098	0.3168 0.6736 0.0083 0.3740
Sodium metabisulfite	2.0089 2.0045 2.0067	0.5546 0.2112
Potassium metabisulfite	2.0130 2.0093 2.0092 2.0052 2.0074 2.0078	0.9715 0.2011 0.2404 0.0961

Despite the spectral appearance of isotropic lines of the radical of sodium sulfite, sodium bisulfite and sodium metabisulfite, the spectral simulation details orthorhombic symmetry. A similar result is reported by Gustafsson et al. [14] that found axial symmetry for an SO_3^- radical in irradiated potassium dithionite. On the other hand, Chanty et al. [15] describe an isotropic line for an SO_3^- radical in irradiated sodium dithionite, which leads us to conclude that although the nature of the radical is the same, its spectrum depends on the compound in which it is present.

The power saturation curve is presented in Figure 3, considering the peak-to-peak amplitude of the ESR first derivative absorption line normalized by sample mass.

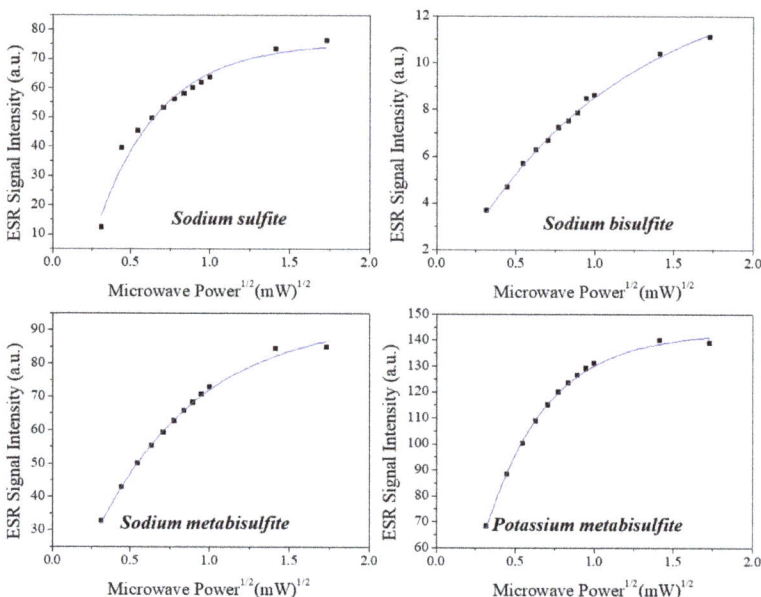

Figure 3. ESR signal amplitude as a function of the microwave incident power at the resonant cavity.

The signal intensity as a function of microwave power for all sulfites demonstrates a conventional characteristic, with a signal that increases with power$^{1/2}$, with subsequent saturation of the signal up to 3 mW, which is a usual characteristic of spin–spin relaxation. The figure shows fitting with a single saturation exponential curve. For comparison purposes, samples with the same dose (500 Gy) were used with the signal normalized by mass. Sodium Bisulfite presented the smallest value of signal intensity at saturation and potassium metabisulfite reached the highest.

The stability of the ESR signal is crucial for retrospective dosimetry. So, the signal intensity as a function of time after irradiation was monitored until 166 h. Figure 4 shows the results, all samples were irradiated with 500 Gy, and the signal was normalized by sample mass for comparison. During this period of study, there was no fading of signals. As already mentioned, the high signal produced with a dose of 500 Gy was valuable as it gave better precision in determining the fading and the microwave power saturation.

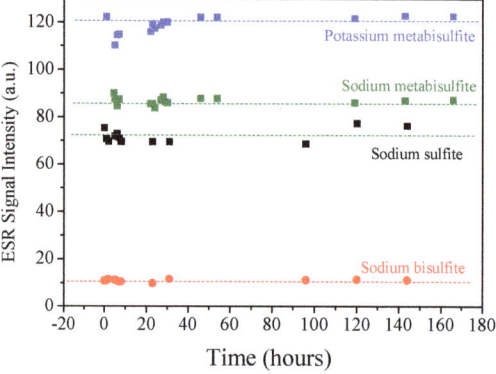

Figure 4. ESR signal amplitude of sample irradiated with a dose of 500 Gy as function of time after irradiation, showing the stability of signals during this period.

2.2. Dose-Response Curve

Table 2 summarizes the best spectrometer parameters for ESR dosimetry with the materials. The modulation amplitude was selected to optimize the signal without its distortion. For sodium sulfite, bisulfite and metabisulfite, a relatively large modulation was employed to increase the signal amplitude. This procedure has been used in other studies [16,17] and showed an alternative when the signal intensity must be correlated with some variable, such as the radiation dose in the present case.

Table 2. X-band ESR spectrometer settings for Dose–Response Curve.

Parameter	Sodium Sulfite	Sodium Bisulfite	Sodium Metabisulfite	Potassium Metabisulfite
Center magnetic field			~337 mT	
Microwave frequency			9 GHz	
Resonant Cavity			Cylindrical resonator, mode TE011, Q factor 6000	
Modulation frequency			100 kHz	
Microwave power			1 mW	
Modulation amplitude	1.0 mT	1.4 mT	1.4 mT	0.1 mT
Sweep width			10 mT	
Time constant			0.3 s	
Gain			10×100	
Sweep time			1 min	
Number of scans			3	
Sample mass	~80 mg	~85 mg	~60 mg	~90 mg

The ESR spectra of the sulfites as a function of the dose are shown in Figure 5. We can observe that the background signal, already present in the materials, has the same structure as the signals induced by ionizing radiation. The amplitude of the background signal was computed in the construction of the Dose–Response curve, ensuring that the calibration of the dosimeter is carried out taking this character into account. We can also notice that the signals increase in intensity with the radiation dose, and, in this dose range, no other species appear. Further experiments are needed, and future work can be performed with DFT to simulate the spectra to identify the nature of the radicals. Three of the compounds exhibit mostly a single ESR line but potassium metabisulfite shows a composite spectrum as demonstrated in the spectral simulation (Figure 2). Thus, the peak-to-peak of the main line was used to construct the Dose–Response curve and a smaller value of field modulation, in comparison to the other compounds, was employed.

Most substances used for ESR dosimetry exhibit exponential behavior over a wide range of doses, up to kGy (Equation (1)). However, for the low dose range, as in the present paper, linear adjustment can be applied, as it is compatible with the beginning of the exponential curve (Equation (2)). Therefore, the experimental data points (Figure 6) were adjusted by linear fitting (Equation (2)), and Table 3 summarizes the parameters of fitting for each sulfite.

$$\mathbf{I} = \mathbf{I_0} \times \left[1 - e^{-\left(\frac{D+\alpha}{D_0}\right)}\right] \quad (1)$$

$$\mathbf{I} = \mathbf{I_0} \times \left(1 + \frac{D}{\alpha}\right) \quad (2)$$

where $\mathbf{I}$ is the signal intensity; $\mathbf{D}$, the dose; $\mathbf{I_0}$, the linear coefficient in Equation (2) and Intensity at saturation in Equation (1); α, the dose related to $\mathbf{I} = 0$; and $\mathbf{D_0}$, the dose at saturation. Table 3 summarizes the parameters of fitting for each sulfite.

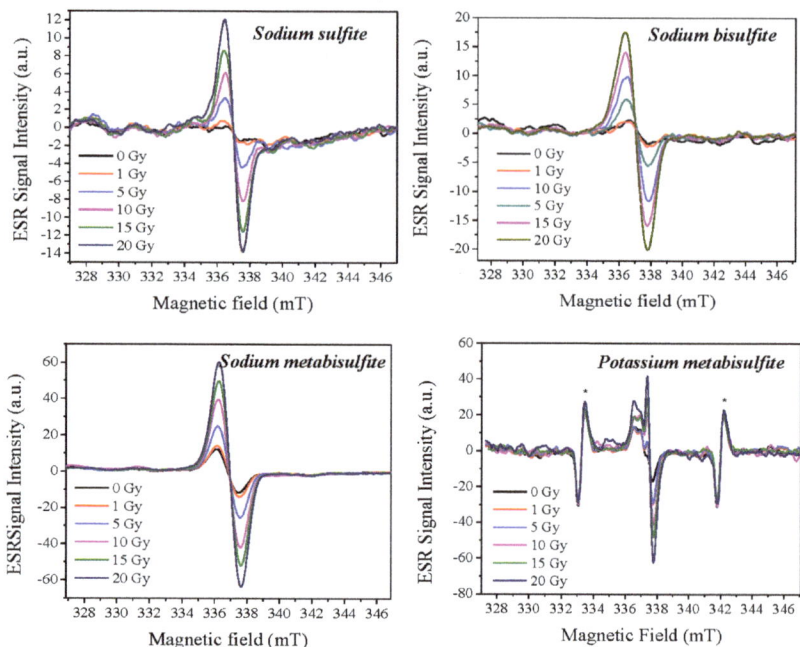

Figure 5. Spectra of sulfites corresponding to each absorbed dose to build the Dose–Response Curve. The 3rd and 4th Mn^{2+} lines of marker are indicated in the Potassium metabisulfite figure (*).

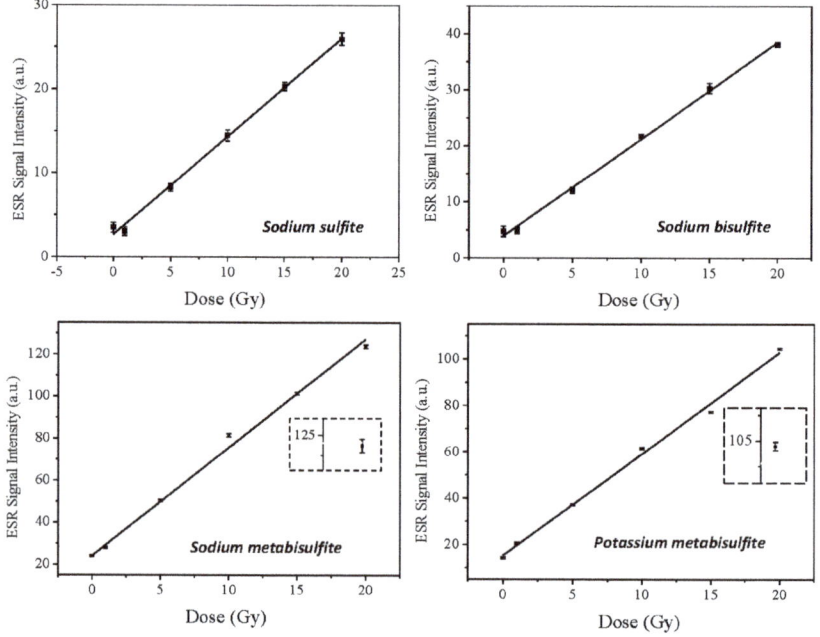

Figure 6. Dose–Response curves of sulfites studied. Linear fitting was used to simplify the data fitting. The inset shows the error box associated with the highest dose for the metabisulfite plots.

Table 3. Parameters of fitting of Dose–Response Curve for each compound studied.

Compound	Model	I_0	α	Adj. R-Square
Sodium sulfite	linear	2.7	2.3	0.997
Sodium bisulfite	linear	4.0	2.3	0.997
Sodium metabisulfite	linear	24.1	4.7	0.997
Potassium metabisulfite	linear	15.3	3.5	0.995

The Dose–Response curves obtained with optimized acquisition parameters for each compound studied demonstrate that metabisulfites are more sensitive to irradiation, because the curves, for the same dose range, reach higher intensities in relation to sodium sulfite and sodium bisulfite. This can be explained by observing that chemical structures composed of more elements show greater intensity of signal, due to the larger quantity of free radicals and their stability. Additionally, tests with radiation doses greater than 20 Gy may confirm the linear behavior of dose–response curves, so that sulfites can be used for high-dose dosimetry.

Therefore, these characteristics open space for future research, in which the behavior can be studied in relation to higher dose ranges, as well as to other energy ranges, opening perspectives for their applications.

3. Materials and Methods

Potassium metabisulfite (Synth), sodium bisulfite (Sigma-Aldrich), sodium metabisulfite (Reagen) and sodium sulfite (Nuclear) of analytical grade were all commercially obtained in powder form and used as received. The samples were not crushed or sieved to avoid creation of defects by mechanical action.

3.1. Irradiation

For irradiation, each material was placed in a small capsule and positioned between a solid water slab with thickness of 1.5 cm on top, corresponding to the build-up region for 6 MV energy, and a 15 cm solid water slab below, to allow proper back scattering conditions. The irradiation was performed with a Siemens Mevatron 6 MV clinical linear accelerator (Linac) with dose values varying from 1 to 20 Gy, for dose response investigation. This dose range was chosen with the aim of using these compounds as fortuitous dosimeters. The samples were positioned at source surface distance (SSD) of 100 cm at the Linac isocenter and irradiated with a 2 Gy/min dose rate in a 10×10 cm^2 square field size. An aliquot of each material was irradiated with a dose of 500 Gy through Cesio-137 source for determination of the spectrum components by computer simulation, signal stability experiments and characterization in relation to microwave power. Such a high dose was employed to fully reveal all possible radicals created by ionizing radiation allowing a more precise spectral simulation and identification of all the radicals.

3.2. ESR Spectra Characterization

The spectra were recorded in a JEOL–JES FA 200 X-band spectrometer, with cylindrical resonator, mode TE$_{011}$. A Mn^{2+} marker was recorded simultaneously with the sample allowing the determination of the g-factor of each sulfite with JEOL software.

The sample was placed in 4 mm diameter high-purity quartz tubes for ESR measurements. Initially, the spectra of the sample before and after irradiation with 500 Gy were recorded for spectral characterization such as determination of the g-factors of radicals. Then, the signal intensity, considered as the peak-to-peak amplitude of the main line obtained by the first derivative of the absorption spectrum normalized by the mass quantity, was monitored as function of microwave power. The value of power that resulted in the higher intensity, without signal saturation, was adopted to construct the Dose–Response curve. Moreover, the signal was monitored over time to study its stability.

To confirm the assignment of the induced radicals by radiation, spectra simulation with the software EasySpin [13] was compared with experimental results.

3.3. Dose-Response Curve

For Dose-Response Curve construction, the peak-to-peak amplitude of the main signal normalized by sample mass as function of dose was used. The software Origin (OriginLab) was used to adjust the experimental data with the linear or exponential function.

4. Conclusions

Sulfites studied in this work presented a proportional response with radiation and stable ESR signal, enabling their use as a dosimeter. The materials used in this study could be good candidates for accidental dosimetry studies since their presence in many food and pharmaceutical factories makes their use possible in the case of a radiological accident. Further studies are required to determine the lowest detectable dose, identification of radicals and influence of the energy of the radiation source among others, to establish a complete protocol for their use.

Author Contributions: Conceptualization, O.R.N. and O.B.; Formal analysis, A.K., O.R.N. and O.B.; Investigation, A.B.R., A.K., P.M.D., O.R.N. and O.B.; Methodology, A.B.R.; Writing—original draft, A.B.R., A.K. and O.B.; Writing—review and editing, A.K., O.R.N. and O.B. All authors have read and agreed to the published version of the manuscript.

Funding: Fundação de Amparo à Pesquisa do Estado de São Paulo (FAPESP) grants 2013/03258-9, 07/06720-4 and 2013/07699-0 and CNPq grants 302186/2019-0, 304107/2019-0 and 309186/2020-0.

Institutional Review Board Statement: Not applicable.

Informed Consent Statement: Not applicable.

Data Availability Statement: The data can be made available by the authors upon request.

Acknowledgments: The authors thank Francisco Sampaio for sample irradiation.

Conflicts of Interest: The authors declare no conflict of interest.

Sample Availability: Samples of the all compounds studied in this paper are available from the authors.

References

1. Gould, G.W.; Russell, N.J. Sulfite. In *Food Preservatives*; Russell, N.J., Gould, G.W., Eds.; Springer: Boston, MA, USA, 2003; pp. 85–101. ISBN 978-0-387-30042-9.
2. Garcia-fuentes, A.R.; Wirtz, S.; Vos, E.; Verhagen, H. Short Review of Sulphites as Food Additives. *Eur. J. Nutr. Food Saf.* **2015**, *5*, 113–120. [CrossRef]
3. Gancheva, V.; Sagstuen, E.; Yordanov, N.D. Study on the EPR/dosimetric properties of some substituted alanines. *Radiat. Phys. Chem.* **2006**, *75*, 329–335. [CrossRef]
4. Ikeya, M. *New Applications of Electron Paramagnetic Resonance: Dating, Dosimetry and Microscopy*; World Scientific: Singapore, 1993; p. 500.
5. Ikeya, M.; Fillho, O.B.; Mascarenhas, S. ESR dating of cave deposits from Akiyoshi-do Cave in Japan and Diabo Cavern in Brazil. *J. Speleol. Soc. Jpn.* **1984**, *9*, 58–67.
6. Kinoshita, A.; Brunetti, A.; Avelar, W.E.P.; Mantelatto, F.L.M.; Simões, M.G.; Fransozo, A.; Baffa, O. ESR dating of a subfossil shell from Couve Island, Ubatuba, Brazil. *Appl. Radiat. Isot.* **2002**, *57*, 497–500. [CrossRef]
7. Küçükuysal, C.; Engin, B.; Türkmenoğlu, A.G.; Aydaş, C. ESR dating of calcrete nodules from Bala, Ankara (Turkey): Preliminary results. *Appl. Radiat. Isot.* **2011**, *69*, 492–499. [CrossRef] [PubMed]
8. Tsang, M.-Y.; Toyoda, S.; Tomita, M.; Yamamoto, Y. Thermal stability and closure temperature of barite for electron spin resonance dating. *Quat. Geochronol.* **2022**, *71*, 101332. [CrossRef]
9. Bogushevich, S.; Ugolev, I. Inorganic EPR dosimeter for medical radiology. *Appl. Radiat. Isot.* **2000**, *52*, 1217–1219. [CrossRef]
10. Baran, M.P.; Bugay, O.A.; Kolesnik, S.P.; Maksimenko, V.M.; Teslenko, V.V.; Petrenko, T.L.; Desrosiers, M.F. Barium dithionate as an EPR dosemeter. *Radiat. Prot. Dosim.* **2006**, *120*, 202–204. [CrossRef] [PubMed]
11. Danilczuk, M.; Gustafsson, H.; Sastry, M.D.; Lund, E.; Lund, A. Ammonium dithionate—A new material for highly sensitive EPR dosimetry. *Spectrochim. Acta Part A Mol. Biomol. Spectrosc.* **2008**, *69*, 18–21. [CrossRef] [PubMed]
12. Murthy, G.S.; Eager, R.L.; McCallum, K.J. Radiation Chemistry of Dithionates. *Can. J. Chem.* **1971**, *49*, 3733–3738. [CrossRef]
13. Stoll, S.; Schweiger, A. EasySpin, a comprehensive software package for spectral simulation and analysis in EPR. *J. Magn. Reson.* **2006**, *178*, 42–55. [CrossRef] [PubMed]

14. Gustafsson, H.; Lund, A.; Hole, E.O.; Sagstuen, E. SO$_3^-$ radicals for EPR dosimetry: X- and Q band EPR study and LET dependency of crystalline potassium dithionate. *Radiat. Meas.* **2013**, *59*, 123–128. [CrossRef]
15. Chantry, G.W.; Horsfield, A.; Morton, J.R.; Rowlands, J.R.; Whiffen, D.H. The optical and electron resonance spectra of SO$_3^-$. *Mol. Phys.* **1962**, *5*, 233–239. [CrossRef]
16. Gallo, S.; Iacoviello, G.; Bartolotta, A.; Dondi, D.; Panzeca, S.; Marrale, M. ESR dosimeter material properties of phenols compound exposed to radiotherapeutic electron beams. *Nucl. Instrum. Methods Phys. Res. Sect. B Beam Interact. Mater. Atoms.* **2017**, *407*, 110–117. [CrossRef]
17. Longo, A.; Basile, S.; Brai, M.; Marrale, M.; Tranchina, L. ESR response of watch glasses to proton beams. *Nucl. Instrum. Methods Phys. Res. Sect. B Beam Interact. Mater. Atoms.* **2010**, *268*, 2712–2718. [CrossRef]

Article

Application of Amino Acids for High-Dosage Measurements with Electron Paramagnetic Resonance Spectroscopy

Yordanka Karakirova

Institute of Catalysis, Bulgarian Academy of Sciences, 1113 Sofia, Bulgaria; daniepr@ic.bas.bg

Abstract: A comparative investigation of amino acids (proline, cysteine, and alanine) as dosimetric materials using electron paramagnetic resonance (EPR) spectroscopy in the absorbed dosage range of 1–25 kGy is presented. There were no signals in the EPR spectra of the samples before irradiation. After irradiation, the complex spectra were recorded. These results showed that the investigated amino acids were sensitive to radiation. In the EPR spectrum of cysteine after irradiation, RS• radicals dominated. The effects of the microwave power on the saturation of the EPR signals showed the presence of at least three different types of free radicals in proline. It was also found out that the DL-proline and cysteine had stable free radicals after irradiation and represented a linear dosage response up to 10 kGy. On the other hand, the amino acid alanine has been accepted by the International Atomic Energy Agency as a transfer standard dosimetry system. In view of this, the obtained results of the proline and cysteine studies have been compared with those of the alanine studies. The results showed that the amino acids proline and cysteine could be used as alternative dosimetric materials in lieu of alanine in a dosage range of 1–10 kGy of an absorbed dose of γ-rays using EPR spectroscopy. Regarding the radiation sensitivity, the following order of decreased dosage responses was determined: alanine > DL-proline > cysteine > L-proline.

Keywords: electron paramagnetic resonance (EPR) spectroscopy; amino acids; proline; cysteine; alanine; dosimetry; free radicals; γ-radiation

1. Introduction

Among the various methods of dosimetry, for example, polarimetry, photo- and thermo-luminescence, measurements of electroconductivity and dielectric losses, etc., EPR dosimetry has particular significance. In a number of cases, EPR spectroscopy has shown advantages over the other methods. These advantages include a high sensitivity for a wide measurement range with high accuracy, the small size of the used samples, the non-destructive character of the measurements, and the automation of the processing of the dosimetric data. Many scientists have made valuable contributions to the development of reference standard dosimeters for high energy radiation on the basis of using alanine in an EPR dosimetry system [1,2]. Until the present, this has been the most common material used in EPR dosimetry, and it has been formally accepted by International Atomic Energy Agency (IAEA, Vienna, Austria) [3], the National Institute for Standards and Technology (NIST, Maryland, USA) [4], and the National Physical Laboratory (NPL, Teddington, UK) [5] as a secondary reference and transfer dosimeter for high-dosage irradiation. Alanine EPR dosimetry has been applied successfully for measuring intermediate and high radiation doses. Although the performance of alanine dosimetry has improved, the sensitivity of the material is too low for a fast and simple low-dosage determination. However, there are widely spread applications of alanine, and many scientists continue to search for alternative materials with better characteristics. Materials with greater sensitivity are required to make EPR dosimeters competitive with other dosimetry systems. Strategies for identifying new EPR dosimeter materials have been proposed by Ikeya et al. and Lund et al. [6,7]. The criteria that should be fulfilled by a useful dosimetry system can be divided into radiation dosimetry criteria and radiation chemistry criteria in regard to EPR properties. The important radiation dosimetry criteria are tissue equivalence (with respect to scattering) and the energy absorption of ionizing radiation, as well as the stability of radicals over time and the linearity of the signal versus the dose. The radical stability and linearity of the signal with respect to dose must be verified experimentally, and tissue equivalence excludes materials containing heavy elements. The important radiation chemistry criteria are a high radical yield and a suitable radical structure, which provide longer-living radicals with a simple symmetrical line and a short longitudinal relaxation time period. Many studies have already been carried out in an effort to identify new materials for electron paramagnetic resonance dosimetry and to substitute alanine in such dosimetry [7–16]. Many substances, such as saccharides [17], formates [18,19], tartrate [20], dithionate [21], and ascorbic acid [22], have been studied as dosimetric materials, and a number of amino acids in which free-radical populations form during irradiation have been suggested for high-dosage dosimetry using electron paramagnetic resonance analysis. Several compounds, all of which have been found to be more sensitive than alanine by a factor of 2–10, have been investigated [23,24]. Sucrose has also been widely studied as a dosimeter in radiation accidents, for irradiation with different types ionizing radiation, and with mixed types of radiations [17,25–30]. In the current study, the dosimetric properties of the amino acids proline and cysteine were investigated and compared with those of alanine. These materials were chosen because they are known to have good characteristics as dosimetric materials. They belong to a class of biological substances which, after irradiation, exhibit reasonably well-resolved spectra. It is convenient to study them by EPR spectrometry for two reasons: amino acids are components of proteins that can be purified in crystalline form, and they are used in many foods and food additives. Because of this, if they show good results, they could potentially be used for accidental and/or retrospective dosimetry. The application of the amino acids to a successful and versatile free radical method of dosimetry depends on the magnitude of the radical yield per unit of absorbed dose and on the lifetime of the free radicals. On the other hand, proline and cysteine are among the few left amino acids that have not yet been studied with respect their use in EPR dosimetry. In view of this, in the present study, all characteristics for dosimetric materials such as sensitivity to radiation, time stability of the radiation, the created free radicals, and the dose–response characteristics of proline and cysteine were studied. The obtained results showed the possibilities of using these materials for dosimetric purposes for γ radiation, and this study will enrich the existing knowledge about the EPR dosimetry of amino acids.

2. Results and Discussion

2.1. EPR Spectra

No EPR signals were observed in the samples before irradiation. After irradiation, complex EPR spectra were recorded. It is known that complex spectra are composed as a result of the superposition of the signals of several free radicals. The spectra of L- and DL- proline are shown in Figure 1a,b. As can be seen, the EPR spectra of L- and DL-proline are similar but not exactly the same. The EPR spectrum of DL-proline is characterized by a g factor of 2.00378 ± 0.00002 of the central line, a constant of the hyperfine splitting of $A \approx 2.171$, and a linewidth of $\Delta H \approx 0.94$ mT. The EPR spectrum of L-proline is also centered at a g value of 2.00379 ± 0.00002, and its most intensive three lines have widths of $\Delta H \approx 0.96$ mT and a splitting value of $A = 2.107$. As DL-proline is a racemic mixture of the isomers D- and L-proline, it was not expected to have a different EPR spectrum than that of L-proline. The difference was explained based on the type of sample. In comparison with the DL-proline sample, which was crystalline, the L-proline sample was in the form of powder. This supposes a higher hygroscopicity of the sample. It is known that the absorbance of moisture from the air leads to decreases in the quantity of free radicals and the intensity of the EPR signal, respectively, because of recombination processes. Likely, some of the radicals were more sensitive to the moisture and they disappeared because of the recombination, and this change the shape of the spectra at all.

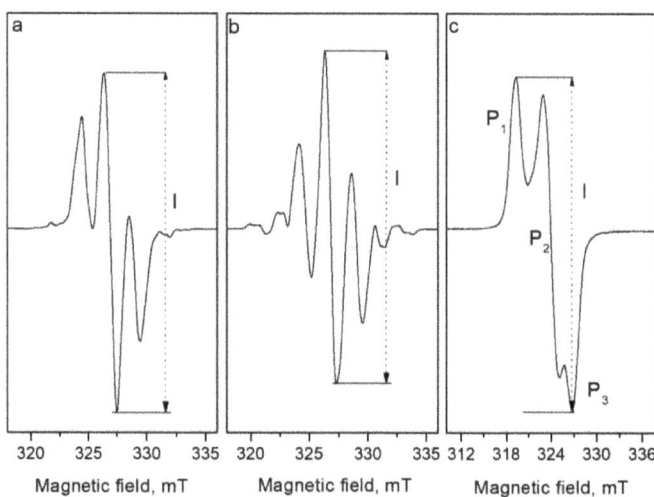

Figure 1. EPR spectra of (**a**) L-proline, (**b**) DL-proline, and (**c**) cysteine after γ-irradiation with 10 kGy.

To compare both spectra, Figure 2 shows the EPR spectra of DL- and L-proline irradiated with 25 kGy. It can be seen that some additional lines are observed at the low and high magnetic fields in the EPR spectra of DL-proline. Besides these extra lines, a small difference in the linewidths of the two lines on the left and on the right of the central line is observed. This can be related to the various relaxation times that characterized the interaction of the electron spin with the surroundings and with each other. Thus, in some cases, the lifetime of the individual spin-orientation state in the radical, or that of the radical itself, may be so short time that the linewidth is affected. These effects can arise from the electron exchange and transfer between molecular species. However, this is not so important for the aim of the current study because there were no observed differences between the spectra of the samples of the L-proline irradiated with different doses and the spectra of the DL-proline samples irradiated with different doses. Figure 3 shows the spectra of the investigated samples irradiated with different doses of gamma rays. The fact that there are differences between the spectra of the various materials did not influence the results with respect to their dosimetric properties because they are being studied as independent dosimetric materials. Since there were no changes in the EPR spectrum with the dosage, an amplitude of the first derivative ("peak-to-peak", from maximum to minimum, denoted by I in Figure 1) can be taken as a relative measure of the quantity of the free radicals. The EPR spectra of cysteine consisted of three lines with the g factors $g_1 = 2.0542$, $g_2 = 2.0251$, and $g_3 = 2.0053$, which are denoted by P1, P2, and P3, respectively, in Figure 1c. According to the literature data, the spectrum of cysteine after irradiation is due to the domination of RS• radicals [31]. This "sulphur pattern" is also found in the spectra of various thiols and in compounds containing S-S bonds after gammairradiation at room temperature. An interaction with one or two methylene protons in $RCH_2S\cdot$ radicals may be observed, though, generally, the proton interactions are too small to further characterize the trapped species. However, a low intensive signal located between P1 and P2 in the spectrum, due to another type of radical with an unknown nature, which was more visible and is denoted by the arrow in Figure 3, was also observed.

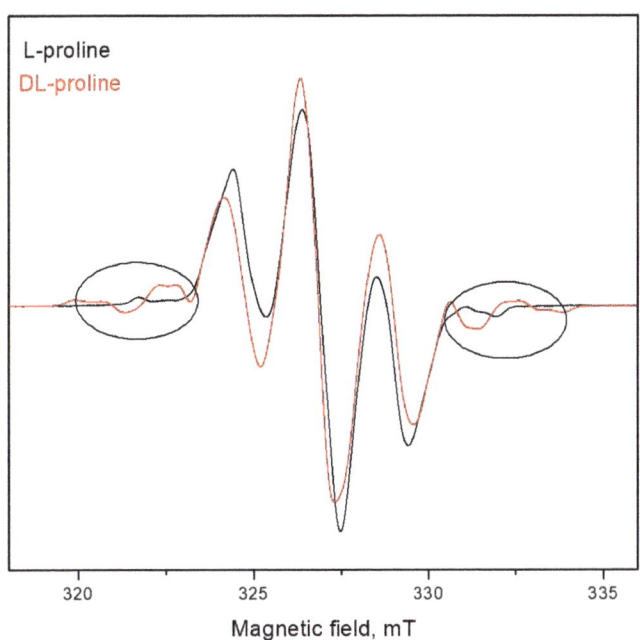

Figure 2. EPR spectra of L- and DL-proline irradiated with 25 kGy.

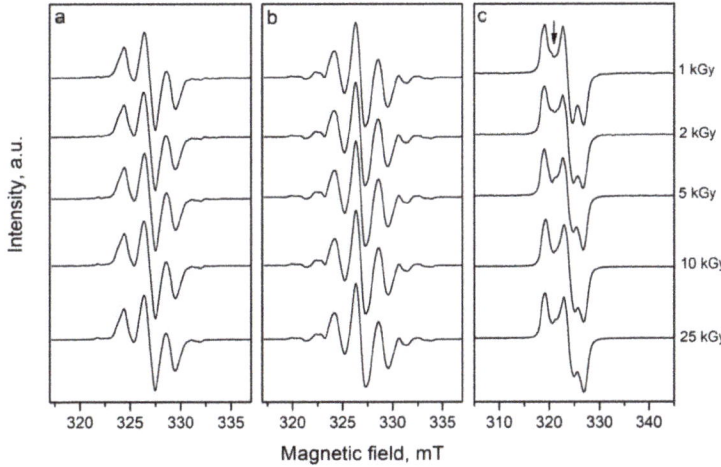

Figure 3. Effect of the dose on the shape of the EPR spectra: (**a**) L-proline, (**b**) DL-proline, and (**c**) cysteine.

The well-known powder EPR spectrum of irradiated α-alanine consists of five broad lines having intensity ratios of 1:4:6:4:1, separated by approximately 2.5 mT (though this is not shown). The observed quintet spectrum is attributed to the hyperfine interaction of the unpaired electron with four protons—three protons from the methyl group and one proton from the α-carbon atom—$CH_3\dot{C}\cdot HCOOH$. The peak-to-peak amplitude of the most intensive central line is commonly used to monitor the dosage deposited in alanine samples after exposure to ionizing radiation. Many studies on the composition of the alanine EPR spectrum are available in the literature [32–35]. It has been shown that the EPR spectrum of irradiated alanine consists of at least three different radical species [36–38].

2.2. Effect of the Dose on the Shape of the EPR Spectra

As mentioned above, Figure 3 shows the effect of the dosage on the shape of the signals in the EPR spectra. It is shown that in the spectra of L- and DL- proline, the different doses do not lead to any changes in the shape of the spectra. At the same time, a small change in the intensity of the central line in the spectrum of cysteine, relative to both lines from right and left, with the increased doses was observed. This can likely be produced by different densities of the radiation beams during irradiation with lower (1 kGy) and higher (25 kGy) doses. However, this may also be due to saturation at high dosages of the radicals responsible for this line. Previous investigations have shown that different dosages do not impact the shape of the spectra of alanine.

2.3. Effect of the Microwave Power on the Shape and on the Saturation Degree of the EPR Signals

The intensities of the EPR signals are known to depend on the values of the instrumental settings, i.e., the microwave power and modulation amplitude. Therefore, the first step after irradiation was to study the influence of these parameters on the EPR response. In view of this fact, two series of investigations on the dependence of EPR intensity as a function of the square root of the microwave power and of the magnetic field modulation amplitude were made. The results (Figures 4 and 5) showed that for proline, the EPR intensity remained linearly dependent on the microwave power up to 0.3 mW and on the modulation amplitude up to 0.4 mT. For cysteine, the sample peaks 1 and 3 had linear dependence up to 6 mW, whereas peak 2 was saturated at a lower value of the microwave power (1 mW). The dependencies on the modulation amplitude were linear up to 0.4 mT. The values of the parameters that were chosen for the measurements were required be in the linear parts of the graphs. However, the appropriate instrumental settings to record the EPR spectra of alanine were previously studied to compare the spectra, and it was acceptable to determine them using the same instrument. Therefore, the following values of the parameters were identified: in the case of proline, a microwave power 0.3 mW and a modulation amplitude 0.4 mT were applied, and for cysteine and alanine, microwave power of 1 mW and modulation amplitude of 0.4 mT, respectively, were applied.

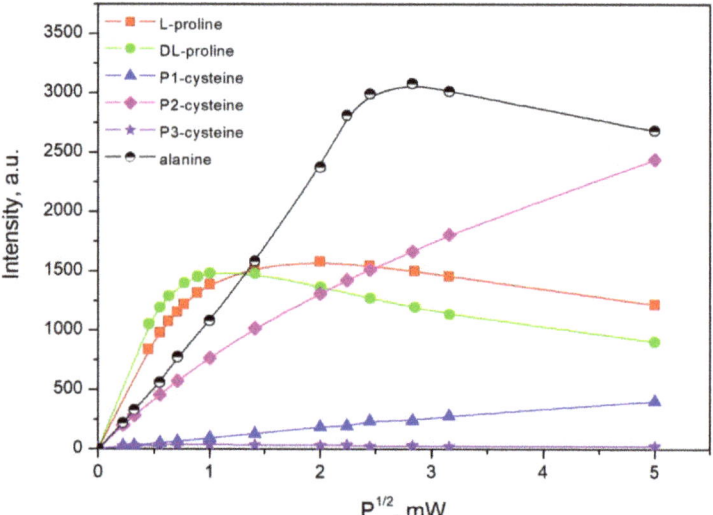

Figure 4. The dependence of the EPR intensity on the square root of the microwave power.

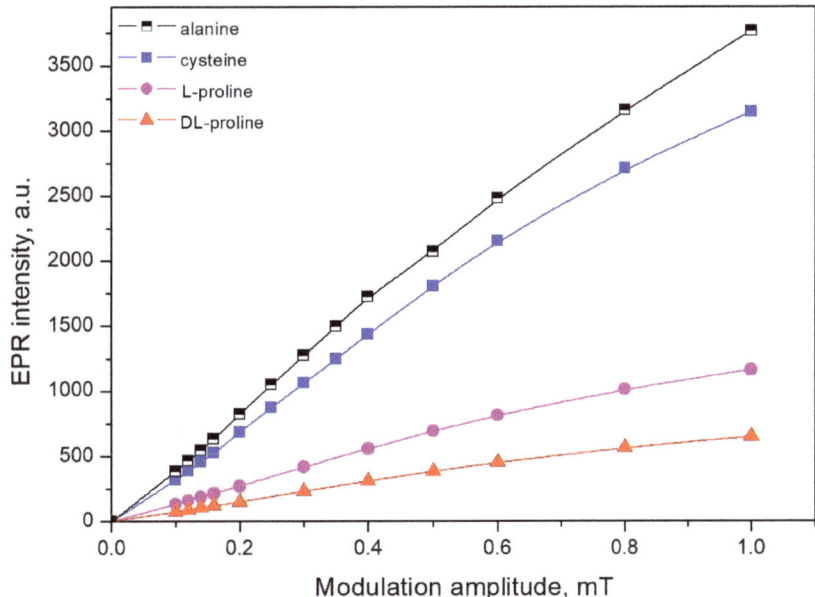

Figure 5. The dependence of the EPR signal intensity on the modulation amplitude.

Figure 6 shows the behavior of the saturation of the EPR signal of the observed lines in the spectra of proline with regard to the microwave power. For this study, a sample of DL-proline irradiated with 5 kGy was used. The results for L-proline were the same, which is why they are not shown. The number of each line can be seen on the left panel in the figure, whereas on the right, the dependence on the square root of the microwave power is shown. As seen in the figure, peaks 3, 4, and 5 have similar behaviors, namely, a linear dependence up to 0.3 mW, which slowly decreased after that point. Peaks 2 and 6 decreased with the increase in microwave power. When the power exceeded 1 mW, peak 6 disappeared. The changes in the magnitude of the microwave power weakly influenced the intensity of peaks 1 and 7. At values higher than 1 mW, peak 1 was not observed. On the basis of these results, it can be concluded that at least three types of free radicals were created in proline during the γ-irradiation. The first one was responsible for peaks 3, 4, and 5 in the EPR spectra of proline. Peaks 2 and 6 in the spectra are due to the second type of radical. The last radical was responsible for peaks 1 and 7. Similar to this, if we look at the dependence of different peaks in the spectra of cysteine with regard to the microwave power (Figure 4, P1, P2, and P3), it can be seen that two different saturation behaviors were observed. One of them was for peaks P1 and P3 and the second was for P2. Therefore, this is evidence that in addition to RS• radicals in the EPR spectra, there are also contributions by other paramagnetic species with unknown nature. This statement is in accordance with the observation in Figure 3, where it can be seen that P2 had changed its intensity regarding P1 and P3 after the increase in the dose.

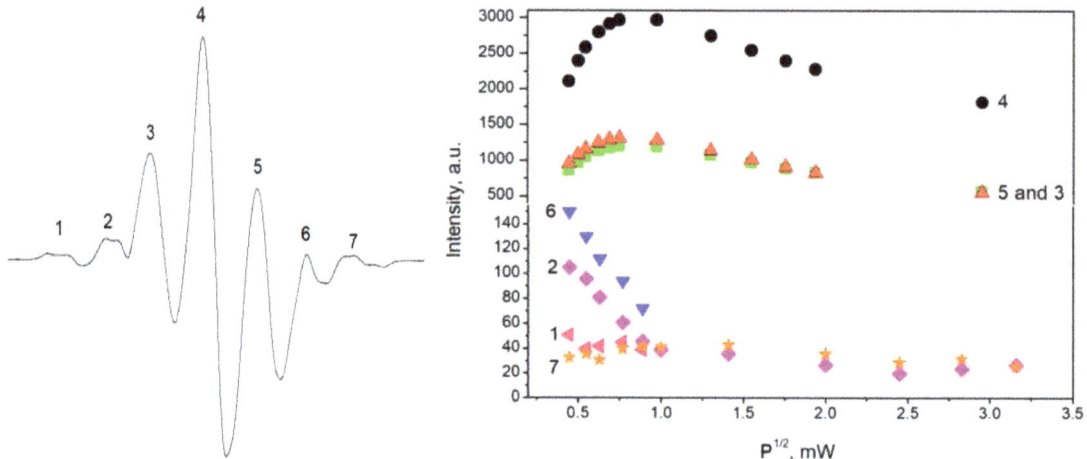

Figure 6. Saturation behaviors of the EPR lines in the spectra of proline versus the square root of the microwave power.

The investigation of the saturation effect of the lines in the EPR spectra of alanine upon increase in the microwave power showed the presence of three different types of radicals (R1, R2, and R3) [36].

2.4. Time Dependence of the Free Radicals

The time stability of free radicals depends on their molecular surroundings and, especially, on the state of the atom. Normally, in solutions, the lifetimes of the unpaired electrons or free radicals are very short. Some crystalline materials, even at room temperature, can exist for as long as several years. The type and amount of the free radicals created by ionizing radiation depend on the crystalline structure and storage conditions.

From a dosimetry point of view, the knowledge of the time stability of the radiation-induced EPR signal in the samples, as well as their decay kinetics, are highly important. This is especially important when several days can elapse between the exposure and the instrumental evaluations. There are at least two mechanisms of decay in the monitored free radicals: recombination with other paramagnetic species to create diamagnetic products and their transformation into another paramagnetic molecule. In the first case, only the intensity of the obtained EPR spectrum will decrease with time and no changes in its shape will be observed, and in the second case, new EPR spectra will appear. As it is typical, the effects of the second type were observed immediately after irradiation of the substance under study until stable paramagnetic species were formed. After that, the remaining stable free radicals could only recombine. In view of this, all measurements were performed at least 72 h after irradiation in order to avoid short-living intermediate relaxing products.

For this study, the samples were stored at room temperature in the dark and then measured for a period of six months. The results showed that for this period, the intensity of the DL-proline decreased by approximately 25% (Figure 7a). The radiation-induced signal of L-proline decreased by 83% for 3 months (Figure 7b). Six months after irradiation, the signals had decreased by 99% and nearly disappeared. This result can be explained by the fact that the samples of L-proline are more hygroscopic than DL-proline. The L-proline was in the form of powder, whereas the DL-proline samples were crystals. However, the samples were stored under the same conditions, it was visible that the samples of L-proline had absorbed moisture from the air, even though they were stored in plastic bags in a dry and dark place. In the results, the recombination of the free radicals was observed, and therefore, there was a decrease in the signal intensity.

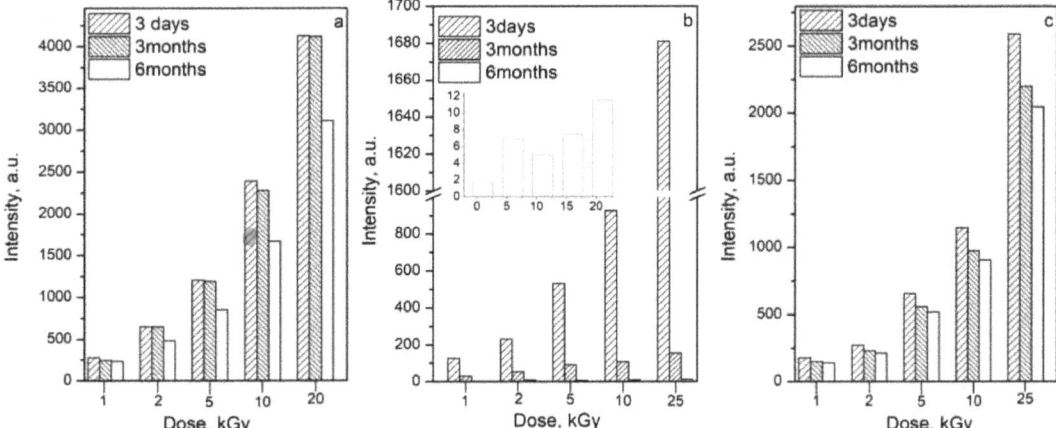

Figure 7. Time stability of the free radicals recorded for: (**a**) DL-proline; (**b**) L-proline, where the small graph shows the EPR signal intensity after six months of storage; and (**c**) cysteine.

The cysteine intensity decreased by 21% over the same time period of six months (Figure 7c).

The investigations of the lifetimes of the radiation-induced radicals in alanine were completed more than 35 years ago. They showed that the lifetimes of the free radicals were very long [39]. It was reported that the decay rate was less than 1% for 3 years. On the other hand, Hansen and Olsen [40] found a dependence between radical decay and applied dosage. They discovered very low fading for doses below 10 kGy and more pronounced fading for doses above 50 kGy. This showed that the radicals could likely begin to interact with each other above a certain concentration. This fact can also explain the saturation of the signal intensity at higher doses. There have been many other studies on the time stability of irradiation-induced free radicals in alanine, and as a whole, they concluded that they were stable for a long time period.

2.5. Dose–Response Characteristics

The dose–response characteristics of proline, cysteine, and alanine were obtained for ^{137}Cs γ-rays. The responses were expressed as changes in the EPR signal intensity ("peak-to-peak" amplitude of the first derivative, denoted by "I" in Figure 1) of the irradiated samples as a function of the absorbed dose. The dose–response curves are shown in Figure 8. Each data point consists of three independent measurements of three separate samples that were simultaneously irradiated.

The results showed the linear dependence of the EPR signal intensity on the absorbed dose gamma rays up to 10 kGy and the saturation of the intensity at 25 kGy. These results are in accordance with those published in the literature data for the dose–response characteristics of other materials, for example, those of mono- and di-saccharides [6]. With respect to radiation sensitivity, the following order of decreases in sensitivity was determined: alanine > DL-proline > cysteine > L-proline. The dose–response curves were built with the data obtained 72 h after irradiation. This was necessary to avoid the short-living intermediate relaxing products in the first hours after radiation treatment.

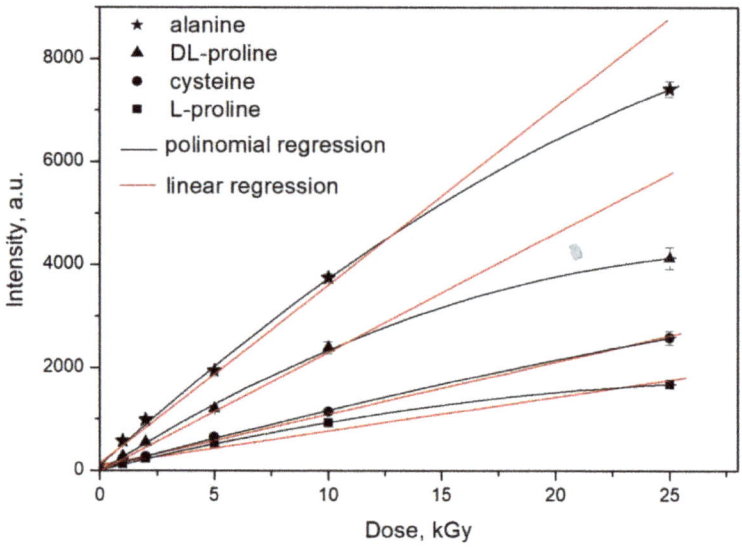

Figure 8. Dependence of the EPR signal intensity on the applied dose of gamma radiation.

Because of the saturation processes, two types of equations were used to fit the results: linear and polynomial regression. The obtained data are shown in Table 1.

Table 1. Coefficient values for the dependencies of the EPR signal intensity on the applied dosage.

Sample	Linear Regression $I = A + BD$			Polynomial Regression $I = A + B_1 + B_2D^2$			
	A	B	R	A	B1	B2	R
DL-proline	251.45	163.81	0.9872	17.93	279.50	−4.60	0.9984
L-proline	110.58	65.94	0.9865	14.32	109.46	−1.71	0.9997
Cysteine	77.41	101.76	0.9984	33.32	121.70	−0.78	0.9992
Alanine	360.99	290.95	0.9944	97.55	410.05	−4.69	0.9991

I—intensity of the EPR signal; A, B—coefficients; D—the absorbed dose of radiation in kGy; R—correlation coefficient.

3. Materials and Methods

3.1. Materials

The amino acids (L-proline, DL-proline, alanine, and cysteine) were bought from Sigma Aldrich. Both forms of proline—L- and DL—were used. The chemical structure of cysteine and the isomers of proline are shown in Scheme 1. The structure of alanine is not shown because it is well known.

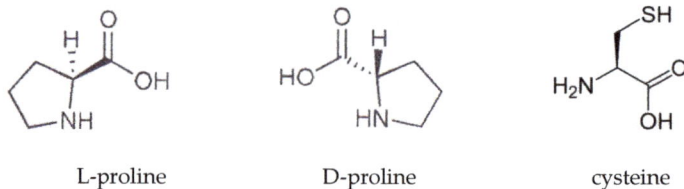

L-proline   D-proline   cysteine

Scheme 1. Chemical structures of proline and cysteine.

L-proline is the only proteinogenic amino acid that is a secondary amine, meaning that its amine nitrogen is bound to two alkyl groups. It is especially important in the production of collagen, which is a primary component in skin, cartilage, and bones. L-proline can be found in a large number of food supplements to support the growth of connective tissue. DL-proline is a racemic mixture of the naturally occurring isomers of L and D-proline. D-amino acids have been found in relatively high abundance in human plasma and saliva.

Cysteine is a sulfur-containing, semi-essential proteinogenic amino acid. It can be synthesized by the human body under normal physiological conditions if a sufficient quantity of methionine is available. The majority of L-cysteine is obtained industrially through the hydrolysis of animal materials, such as poultry feathers or hog hair. Cysteine, primarily the L-enantiomer, is a precursor in the food, pharmaceutical, and personal care industries. One of its largest applications is in the production of flavors.

Alanine is the simplest α-amino acid after glycine. The methyl side-chain of alanine is non-reactive, and therefore, it is rarely directly involved in protein function. Alanine is a nonessential amino acid, meaning it can be manufactured by the human body, and it does not need to be obtained through the diet. Alanine is found in a wide variety of foods, but it is particularly concentrated in meats.

3.2. Irradiation

Three parallel samples of each amino acid were taken and irradiated with gamma rays. The irradiation was performed by applying a source, ^{137}Cs, a dosage rate of 200 Gy/h, and a dosage range of 1–25 kGy. For control of the absorbed dose distribution, we used the alanine dosimeters of a Kodak BioMax. Three dosimeters were placed at each point. The control measurements and calibration of the absorbed dosage in water were completed by an X-band EPR spectrometer (E-scan, Bruker). The irradiation was performed in air and at room temperature. After irradiation, all the samples were kept in closed plastic bags at room temperature and stored in the dark.

3.3. Principles of the EPR Method

EPR spectroscopy can be defined as the resonant absorption of electromagnetic energy in paramagnetic substances by the transition of the spin of an unpaired electron between different energy levels (a state of lower energy and a higher-energy state), in the presence of a magnetic field. In the presence of an external magnetic field, the spin of unpaired electron is orientated to it in two directions: parallel and antiparallel to the field. The energy difference, ΔE, between these levels is proportional to the Lande g-factor, the Bohr magneton, β, and the magnetic field, H. The relationship is given by the equation:

$$\Delta E = g\beta H$$

In case of thermal equilibrium, the population of the lower energy level E_1, is slightly higher than that of the upper level, E_2. Therefore, the system is able to absorb energy, $\Delta E = h\nu$, from an external high-frequency field. When the sample is irradiated using radiation with an appropriate frequency and $h\nu = g\beta H$, transitions from the lower to the upper state appear and the EPR spectrum, as a first derivative of the absorption curve, is recorded.

3.4. Instrument

The EPR spectra were recorded using a JEOL JES FA 100 EPR spectrometer at room temperature. The X-band EPR spectrometer was operated at 9.5 GHz of frequency, and it had a standard TE$_{011}$ cylindrical resonator.

3.5. Procedure of Measurement

For each single measurement, an equal weight of the samples was placed in quartz tubes (4 mm inner diameters). For the best sensitivity, the tubes were positioned in the center of the EPR cavity. Three independent measurements were used for every sample, including a procedure where inserting-removing-inserting of the sample was performed in the cavity of the EPR spectrometer. The data were averaged, and in that way, the error of the measurement was determined to be 3%. A reference sample, Mn magnetic diluted in MgO, which is an internal standard in the above-mentioned spectrometer, was analyzed before and after each series of measurements under the same conditions as those used for the sample measurements to normalize the signal intensity of the samples and to minimize the error resulting from any instability in the spectrometer. The parameters for recording the spectra were as follows: modulation frequency of 100 kHz, microwave powers of 0.3 mW (proline) and 1 mW (cysteine and alanine), modulation amplitude of 0.4 mT, time constant of 0.03 s, and sweeping time of 2 min.

4. Conclusions

After irradiating DL-proline, L-proline, and cysteine samples with γ-rays, complex EPR spectra of all samples were recorded. The effect of the microwave power on the shape and the saturation of the EPR signals showed that at least three types of free radicals with unknown natures were created in proline during the γ-irradiation, and two radicals were created in cysteine, one of which was RS•. The time dependence analysis of the EPR spectra after irradiation shows a fading of intensity of DL-proline with 25 %, L-proline – 99% and cysteine – 21% for six months. For comparison, the free radicals created by the radiation in alanine were stable for a longer time period. For all materials, the EPR signal amplitude had a linear dose response up to 10 kGy, and it was saturated at higher doses. Under the same experimental conditions, alanine also showed a linear response up to 10 kGy but with better sensitivity. All these results provide an opportunity for DL-proline and cysteine to be used as dosimetric materials for doses ranging from 1 to 10 kGy, but they have lower sensitivity than alanine. However, in case of emergency dosimetry, if they are present in such a situation, they could be successfully used for dose assessment. For retrospective dosimetry, they are not very suitable because of the decay rates of the radiation-induced free radicals. On the basis of the conducted research, it can be concluded that alanine remains the best candidate for a universal dosimetric material.

Funding: This research work was funded by the Bulgarian National Science Fund, grant number KP-06-N39/12.

Institutional Review Board Statement: Not applicable.

Informed Consent Statement: Not applicable.

Data Availability Statement: The data analyzed in this study are available from the corresponding author upon reasonable request.

Conflicts of Interest: The author declares no conflict of interest.

Sample Availability: Not available.

References

1. Zagorski, Z.P. Dosimetric applications of α-alanine. *J. Radioanal. Nucl. Chem.* **1994**, *187*, 73–78. [CrossRef]
2. Gancheva, V.; Yordanov, N.D.; Callens, F.; Vanhaelewyn, G.; Raffi, J.; Bortolin, E.; Onori, S.; Malinen, E.; Sagstuen, E.; Fabisiak, S.; et al. An international intercomparison on "self-calibrated" alanine EPR dosimeters. *Radiat. Phys. Chem.* **2008**, *77*, 357–364. [CrossRef]
3. Mehta, K.; Girzikowsky, R. Alanine-ESR dosimetry for radiotherapy IAEA experience. *Appl. Radiat. Isot.* **1996**, *47*, 1189–1191. [CrossRef]
4. Desrosiers, M. Alanine dosimetry at the NIST. In *Book of Abstracts, Proceeding of the International Conference on Biodosimetry and 5th International Symposium on ESR Dosimetry and Applications, Moscow/Obninsk, Russia, 22–26 June 1998*; Publisher: Obninsk, Russia, 1998; p. 149.
5. Sharpe, H.G.; Sephton, J.P. *Alanine Dosimetry at NPL—The Development of a Mailed Reference Dosimetry Service at Radiotherapy Dose Level, in Techniques for High Dose Dosimetry in Industry, Agriculture and Medicine, Proceedings of the Symposium Held in Vienna, Austria, 2–5 November 1999*; International Atomic Energy Agency: Vienna, Austria, 1999; p. 299.
6. Ikeya, M.; Hassan, G.M.; Sasaoka, H.; Kinoshita, Y.; Takakiand, S.; Yamanaka, C. Strategy for finding new materials for ESR dosimeters. *Appl. Radiat. Isot.* **2000**, *52*, 1209–1215. [CrossRef] [PubMed]

7. Lund, A.; Olsson, S.; Bonora, M.; Lund, E.; Gustafsson, H. New materials for ESR dosimetry. *Spectrochem. Acta A* **2002**, *58*, 1301–1311. [CrossRef] [PubMed]
8. Hassan, G.M.; Ikeya, M. Metal ion-organic compound for high sensitive ESR dosimetry. *Appl. Radiat. Isot.* **2000**, *52*, 1247–1254. [CrossRef]
9. Lund, E.; Gustafsson, H.; Danilczuk, M.; Sastry, M.D.; Lund, A.; Vestad, T.A.; Malinen, E.; Hole, E.O.; Sagstuen, E. Formates and dithionates: Sensitive EPR-dosimeter materials for radiation therapy. *Appl. Radiat. Isot.* **2005**, *62*, 317–324. [CrossRef]
10. Olsson, S.; Sagstuen, E.; Bonora, M.; Lund, A. EPR dosimetric properties of 2-methylalanine: EPR, ENDOR and FT-EPR investigations. *Radiat. Res.* **2002**, *157*, 113–121. [CrossRef]
11. Gancheva, V.; Sagstuen, E.; Yordanov, N.D. Study on the EPR/dosimetric properties of some substituted alanines. *Radiat. Phys. Chem.* **2006**, *75*, 329–335. [CrossRef]
12. Soliman, Y.S.; Abdel-Fattah, A.A. Magnesium lactate mixed with EVA polymer/paraffin as an EPR dosimeter for radiation processing application. *Radiat. Phys. Chem.* **2012**, *81*, 1910–1916. [CrossRef]
13. Lelie, S.; Hole, E.O.; Duchateau, M.; Schroeyers, W.; Schreurs, S.; Verellen, D. The investigation of lithium formate hydrate, sodium dithionate and N-methyl taurine as clinical EPR dosimeters. *Radiat. Meas.* **2013**, *59*, 218–224. [CrossRef]
14. Rushdi, M.A.H.; Abdel-Fattah, A.A.; Sherif, M.M.; Soliman, Y.S.; Mansour, A. Strontium sulfate as an EPR dosimeter for radiation technology application. *Radiat. Phys. Chem.* **2015**, *106*, 130–135. [CrossRef]
15. Rushdi, M.A.H.; Abdel-Fattah, A.A.; Soliman, Y. Radiation-induced defects in strontium carbonate rod for EPR dosimetry applications. *Radiat. Phys. Chem.* **2017**, *131*, 1–6. [CrossRef]
16. Gallo, S.; Iacoviello, G.; Bartolotta, A.; Dondi, D.; Panzeca, S.; Marrale, M. ESR dosimeter material properties of phenols compound exposed to radiotherapeutic electron beams. *Nucl. Instrum. Methods Phys. Res. Sect. B Beam Interact. Mater. At.* **2017**, *407*, 110–11715. [CrossRef]
17. Karakirova, Y.; Yordanov, N.D.; De Cooman, H.; Vrielinck, H.; Callens, F. Dosimetric characteristics of different types of saccharides: An EPR and UVspectrometric study. *Radiat. Phys. Chem.* **2010**, *79*, 654–659. [CrossRef]
18. Gustafsson, H.; Danilczuk, M.; Sastry, M.D.; Lund, A.; Lund, E. Enhanced sensitivity of lithium dithionates doped with rhodium and nickel for EPR dosimetry. *Spectrochim. Acta A Mol. Biomol. Spectrosc.* **2005**, *62*, 614–620. [CrossRef]
19. Belahmara, A.; Mikou, M.; El Ghalmi, M. Analysis by EPR measurements and spectral deconvolution of the dosimetric properties of lithium formate monohydrate. *Nucl. Instrum. Methods Phys. Res. Sect. B Beam Interact. Mater. At.* **2018**, *431*, 19–24. [CrossRef]
20. Olsson, S.K.; Lund, E.; Lund, A. Development of ammonium tartrate as an ESR dosimeter material for clinical purposes. *Appl. Radiat. Isot.* **2000**, *52*, 1235–1241. [CrossRef]
21. Danilczuk, M.; Gustafsson, H.; Sastry, M.D.; Lund, E.; Lund, A. Ammonium dithionate—A new material for highly sensitive EPR dosimetry. *Spectrochim. Acta A Mol. Biomol. Spectrosc.* **2008**, *69*, 18–21. [CrossRef]
22. Tuner, H.; Korkmaz, M. ESR study of ascorbic acid irradiated with gamma-rays. *J. Radioanal. Nucl. Chem.* **2007**, *273*, 609–614. [CrossRef]
23. Olsson, S.K.; Bagherian, S.; Lund, E.; Carlsson, G.A.; Lund, A. Ammonium tartrate as an ESR dosimeter material. *Appl. Radiat. Isot.* **1999**, *50*, 955–965. [CrossRef]
24. Olsson, S.; Lund, E.; Erickson, R. Dose response and fading characteristics of an alanine-agarose gel. *Appl. Radiat. Isot.* **1996**, *47*, 1211–1217. [CrossRef]
25. Flores, C.; Cabrera, E.; Calderon, T.; Munoz, E.; Adem, E.; Hernandez, J.; Boldu, J.; Ovalle, P.; Murrieta, H. ESR and optical absorption studies of gamma and electron-irradiation sugar crystals. *Appl. Radiat. Isot.* **2000**, *52*, 1229–1234. [CrossRef] [PubMed]
26. Karakirova, Y.; Lund, E.; Yordanov, N.D. EPR and UV investigation of sucrose irradiated with nitrogen ions and gamma rays. *Radiat. Meas.* **2008**, *43*, 1337–1342. [CrossRef]
27. Mikou, M.; Ghosne, N.; El Baydaoui, R.; Zirari, Z.; Kuntz, F. Performance characteristics of the EPR dosimetry system with table sugar in radiotherapy applications. *Appl. Radiat. Isot.* **2015**, *99*, 1–4. [CrossRef] [PubMed]
28. Karakirova, Y.; Yordanov, N.D. EPR and UV spectrometry investigation of sucrose irradiated with carbon particles. *Radiat. Meas.* **2010**, *45*, 831–835. [CrossRef]
29. Karakirova, Y.; Nakagawa, K.; Yordanov, N.D. EPR and UV spectroscopic investigations of sucrose irradiated with heavy-ion particles. *Radiat. Meas.* **2010**, *45*, 10–14. [CrossRef]
30. Nakagawa, K.; Hara, H.; Matsumoto, K. C-ion and X-ray-induced sucrose radicals investigated by CW EPR and 9 GHz EPR Imaging. *Bull. Chem. Soc. Jpn.* **2017**, *90*, 30–33. [CrossRef]
31. Ayscough, P.R. *Electron Spin Resonance in Chemistry*; Methuen & Co., Ltd.: London, UK, 1967; pp. 349–350.
32. Sagstuen, E.; Hole, E.O.; Haugedal, S.R.; Nelson, W.H. Alanine radicals: Structure determination by EPR and ENDOR of single crystals X-irradiated at 295 K. *J. Phys. Chem. A* **1997**, *101*, 9763–9772. [CrossRef]
33. Heydari, M.Z.; Malinen, E.; Hole, E.O.; Sagstuen, E. Alanine radicals. 2. The composite polycrystalline alanine EPR spectrum studied by ENDOR, thermal annealing, and spectrum simulations. *J. Phys. Chem. A* **2002**, *106*, 8971–8977. [CrossRef]
34. Ban, F.; Wetmore, S.D.; Boyd, R.J. A Density-Functional Theory Investigation of the Radiation Products of L-α-Alanine. *J. Phys. Chem. A* **1999**, *103*, 4303–4308. [CrossRef]
35. Pauwels, E.; Van Speybroeck, V.; Lahorte, P.; Waroquier, M. Density functional calculations on alanine-derived radicals: Influence of molecular environment on EPR hyperfine coupling constants. *J. Phys. Chem. A* **2001**, *105*, 8794–8804. [CrossRef]

36. Malinen, E.; Heydari, M.Z.; Sagstuen, E.; Hole, E.O. Alanine radicals, Part 3: Properties of the components contributing to the EPR spectrum of X-irradiated alanine dosimeters. *Radiat. Res.* **2003**, *159*, 23–32. [CrossRef] [PubMed]
37. Malinen, E.; Hult, E.A.; Hole, E.O.; Sagstuen, E. Alanine radicals, part 4: Relative amounts of radical species in alanine dosimeters after exposure to 6–19 MeV electrons and 10 kV–15 MV photons. *Radiat. Res.* **2003**, *159*, 149–153. [CrossRef]
38. Jåstad, E.O.; Torheim, T.; Villeneuve, K.M.; Kvaal, K.; Hole, E.O.; Sagstuen, E.; Malinen, E.; Futsaether, C.M. In Quest of the Alanine R3 Radical: Multivariate EPR Spectral Analyses of X-Irradiated Alanine in the Solid State. *J. Phys. Chem. A* **2017**, *121*, 7139–7147. [CrossRef] [PubMed]
39. Regulla, D.F.; Deffner, U. Dosimetry by ESR spectroscopy of alanine. *Appl. Rad. Isot.* **1982**, *33*, 1101–1114. [CrossRef]
40. Hansen, J.W.; Olsen, K.J. Theoretical and experimental radiation effectiveness of the free radical dosimeter alanine to irradiation with heavy charged particles. *Radiat. Res.* **1985**, *104*, 15–27. [CrossRef]

Disclaimer/Publisher's Note: The statements, opinions and data contained in all publications are solely those of the individual author(s) and contributor(s) and not of MDPI and/or the editor(s). MDPI and/or the editor(s) disclaim responsibility for any injury to people or property resulting from any ideas, methods, instructions or products referred to in the content.

Article

Conversion of a Single-Frequency X-Band EPR Spectrometer into a Broadband Multi-Frequency 0.1–18 GHz Instrument for Analysis of Complex Molecular Spin Hamiltonians

Wilfred R. Hagen

Department of Biotechnology, Delft University of Technology, Building 58, Van der Maasweg 9, 2629 HZ Delft, The Netherlands; w.r.hagen@tudelft.nl

Abstract: A broadband EPR spectrometer is an instrument that can be tuned to many microwave frequencies over several octaves. Its purpose is the collection of multi-frequency data, whose global analysis affords interpretation of complex spectra by means of deconvolution of frequency-dependent and frequency-independent interaction terms. Such spectra are commonly encountered, for example, from transition-metal complexes and metalloproteins. In a series of previous papers, I have described the development of broadband EPR spectrometers around a vector network analyzer. The present study reports on my endeavor to start from an existing X-band spectrometer and to reversibly re-build it into a broadband machine, in a quest to drastically reduce design effort, building costs, and operational complexity, thus bringing broadband EPR within easy reach of a wide range of researchers.

Keywords: EPR; ESR; broadband; strip line; wire micro strip; metal complex; metalloprotein; free radicals

1. Introduction

EPR spectra generally change shape with changing microwave frequency, ν, because they are monitors of multiple electron interactions, some of which are linear in ν, while others are independent of ν. Linear interactions are the electronic Zeeman interaction and also the g-strain resulting from molecular conformational distributions. Examples of independent interactions are electron–nuclear hyperfine interactions, zero-field interactions in high-spin systems, and dipolar interactions between paramagnetic centers. Collection of data at multiple frequencies, and their subsequent global analysis, increases our chances for a meaningful spectral interpretation in terms of a unique spin Hamiltonian. A common application of this philosophy is multiple high-frequency EPR for the resolution of g matrices with small anisotropy and for the detection of systems with zero-field splitting significantly greater than the standard X-band quantum of ca 0.3 cm^{-1} [1,2]. A similar, though less common, approach at lower frequencies (of order X-band and below) is indicated for resolution and interpretation of complex patterns of hyperfine and/or dipolar interaction.

Over the last decade, I have developed versions of low-frequency broadband EPR as a practical solution for multi-frequency studies around and below X-band [3–6]. Commercial low-frequency EPR spectrometers have been available previously; however, their application has always been limited due to their single-frequency nature: each additional frequency requires an additional investment of order 200 k€ (or k$), and each step in frequency requires the interchange of single-mode resonators and associated cryogenics. My broadband spectrometers were conceived as single stations with a resonator circuit that is tunable to many different frequencies, without the need to change resonators and cooling systems.

In spite of the obvious advantages of the broadband system, it has yet to show wider distribution, which I understand from discussions with interested colleagues is rooted in perceived fear for excessive costs of construction and for excessive complexity in construction and operation. I have therefore re-thought the concept of broadband EPR hardware, and I have now designed a simple and affordable (<15 k€) conversion kit intended as an extension of commercial standard X-band spectrometers (I have used a Bruker EMXplus) into an easily operable general purpose broadband spectrometer for experiments in the approximate frequency range of 0.1–18 GHz. The kit consists of a few readily commercially available microwave devices (source, circulator, amplifier, power meter, detection diode) and a few items that can be easily constructed at low cost and without special skills (resonator, modulation coils). Assembly of the units into a broadband bridge with SMA cables is a ten-minute job, and so is the subsequent implementation of the bridge into the X-band spectrometer. Open-source software to operate the broadband bridge is provided in the form of a simple, modifiable LabVIEW program.

2. Basic Theory

Since 1947, in EPR spectroscopy, the microwave frequency standard is X-band (circa 9–10 GHz) applied to a single-frequency cavity resonator [7]. 'Standard' here means that for many molecular systems, X-band EPR provides an approximate optimum in signal-to-noise ratio, from a balancing of increased sensitivity with frequency due to the Boltzmann distribution over states, and decreased sensitivity with higher frequency due to intrinsic hardware properties such as the noise characteristics of detectors. In practice, this has led to a situation in which the majority of EPR experiments are conducted with commercial X-band spectrometers, and only when the results are insufficient for unequivocal interpretation, additional experiments are carried out in different frequency bands, whereby each additional frequency typically requires the laborious and costly procedure of employing an additional spectrometer. Broadband EPR spectroscopy represents a different approach in which the single-frequency cavity is abandoned in favor of a resonator that can be tuned to many different frequencies over many octaves, all in a single spectrometer. In practice, it is useful to operate the broadband spectrometer next to a standard X-band machine because the sensitivity of the former is significantly lower than that of the latter, although this difference may be less at frequencies outside the X-band [4,5].

Microwaves can be efficiently (that is, with minimal loss of energy) transmitted in a number of ways. In the standard spectrometer, this is carried out by means of waveguides of specific cross dimensions for the X-band. In the broadband spectrometer, transport is via a combination of coaxial cable and, for the resonator, a specific variant of strip-line technology: wire micro strip circuitry, whose basic geometry is illustrated in Figure 1. Its four design parameters, d (wire diameter), h (height of wire above the ground plate), b (extension of sample above the wire), and ε_R (dielectric constant or relative electric permittivity of the sample), determine its characteristic impedance Z_0 via intermediate parameters w (height of square wire equivalent to round wire of diameter d), ε_{eff} (effective permittivity), and ε_{emb} (embedded permittivity) as follows [4]:

$$w = d/1.1803 \tag{1}$$

$$\varepsilon_{eff} = \left(\frac{\varepsilon_R + 1}{2}\right) + \left(\frac{\varepsilon_R - 1}{2}\right)\sqrt{\frac{w}{w + 12h}} + X \tag{2}$$

$$X = 0.04\left(1 - \frac{w}{h}\right)^2 \text{ for } w/h < 1, \text{ and } X = 0 \text{ for } w/h \geq 1 \tag{3}$$

$$\varepsilon_{emb} = \varepsilon_{eff} e^{-2b/h} + \varepsilon_R \left(1 - e^{-2b/h}\right) \tag{4}$$

$$Z_0 = \frac{337}{2\pi\sqrt{\varepsilon_R}} \cosh^{-1}\left(\frac{2h + d}{d}\right) \tag{5}$$

in which 337 Ω is the characteristic impedance of air. With ε_R given for a particular sample, the parameters d, h, and b are adjusted to afford a characteristic impedance as close as possible to Z_0 = 50 Ω. This is the universally used value for all components in microwave circuitry, to guarantee lossless connection. A simple calculator of Z_0 from ε_R, d, h, and b is given in Supplementary Materials.

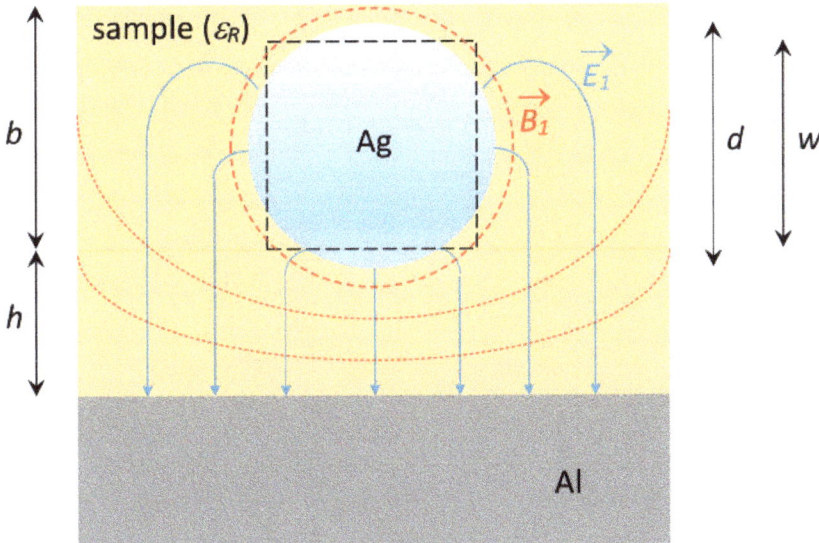

Figure 1. Wire micro strip transmission line for EPR spectroscopy. In this schematic drawing of a cross section through a transmission line, the diamagnetic isolation between a silver wire of diameter d and an aluminium ground plate has been replaced with a paramagnetic sample of height $h + b$ and with electric permittivity ε_R. The square (black broken line), with side length w, is the surface equivalent of the cross-sectional circle of the silver wire. E_1 and B_1 are the vectorial field lines of the electric and magnetic components of the transmitted microwave. The resonator and sample-holder cell to which this cross section belongs is placed in the spectrometer such that the static external magnetic field vector, B_0, is perpendicular to the plain of the drawing, so that all B_1 lines are perpendicular to the B_0 field, as required for EPR. The parameters d (or w), h, b, and ε_R are optimized to obtain a line characteristic impedance $Z_0 \approx 50\ \Omega$.

For example, for a compound with $\varepsilon_R = 3.15$, the values $d = 0.25$ mm, $h = 0.15$ mm, and $b = 0.1$ mm give $Z_0 = 49.7\ \Omega$. Also, the Z_0 value is found to be not very sensitive to the value of b as long as $b > 0$. This leads to a resonator design (geometric details in Section 3) consisting of a metal ground plate covered with four layers of isolating acrylate tape (thickness 50 µm), in which a rectangular sample compartment is cut out that is surrounded by tight windings of metal wire of 0.25 mm diameter such that the isolator is compressed to circa 0.15 mm. When the compartment is filled with sample that extends somewhat above the wire, the resulting characteristic impedance will be approximately 50 Ω. Ice has an $\varepsilon_R = 3.15$ [8], and dilute frozen solutions will not have a very different dielectric constant, so this cell is appropriate, e.g., for frozen aqueous solutions of metal complexes and metalloproteins. Moreover, since powders (microcrystals) of inorganic salts typically have similar ε_R values, broadly in the range 2–5 [9,10], the cell design is also an acceptable compromise for broadband EPR on many inorganics.

3. Results
3.1. Global Description of the Conversion

A schematic overview of the broadband spectrometer is given in Figure 2. Parts of the existing X-band spectrometer are represented in gray; they can be used without modification to control the magnetic field and to encode/decode the EPR signal with 100 kHz field modulation in the signal-channel unit. Also, the proprietary software for operation of the machine is retained (with slightly modified instructions; see below). The X-band bridge is not used and is replaced with a broadband bridge whose elements are represented in pink. All these elements are readily commercially purchasable from multiple vendors. The new bridge is controlled with a dedicated PC (Windows). The X-band cavity (with build-in modulation coils) and waveguide is demounted from the X-band bridge and is removed. It is replaced with items represented in yellow to indicate that they are to be constructed in-house: the multi-frequency resonator (alias sample holder) and a pair of modulation coils providing sufficient space to encompass the resonator and an He/N$_2$ flow dewar. The latter is omitted from the drawing for clarity and is described later.

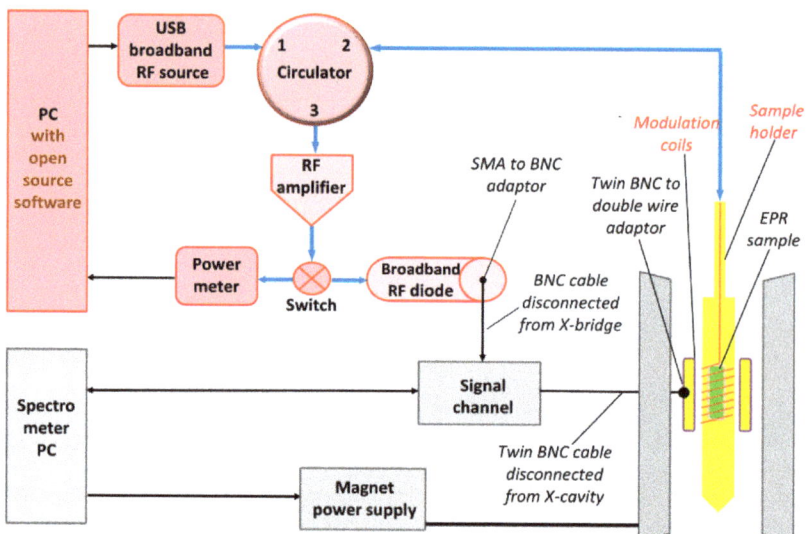

Figure 2. Schematic overview of the proposed broadband EPR spectrometer. An existing, single-frequency, X-band CW spectrometer (parts in gray) with its microwave bridge (not shown) disabled and its cavity removed is extended with a conversion kit consisting of commercially available parts (in pink) and home-made parts (in yellow). Required signal adaptors are also indicated. For clarity, a cooling system is omitted from the drawing.

All microwave, or RF, elements of the bridge are interconnected with blue coaxial cable (e.g., RG402) using SMA male and female connectors. The broadband RF diode detecting the EPR signal has a video output with SMA connector, which is converted to BNC using an adaptor. It is connected to the BNC signal channel cable previously disconnected from the X-bridge, whose other end still connects to the phase-sensitive detector in the signal-channel unit of the X-band spectrometer.

3.2. Details of Purchasable Parts

As a microwave source, I use a SynthHD PRO (ca 3500 €) from Windfreak Technologies (New Port Richey, FL, USA), which is a tunable and scannable source from 10 MHz to 18 GHz with ca +18 dBm maximal output power (the 'm' in 'dBm' defines the power as absolute: 0 dBm corresponds to 1 milliwatt; 18 dBm is 80 mW). Other vendors carry items of similar specifications. The source is controlled by the PC via a USB cable. Note that a microwave source is an active RF device (it consumes energy from a power supply), which means (1) that it produces some heat that must be diverted to a sink such as an aluminium ground plate, and (2) that it is a static-sensitive device, which must be handled with care (ground operator hands) when being connected to other RF circuitry.

The circulator accepts microwaves at its port-1, transfers them to port-2 for delivery to the sample, and then power reflected back from the resonator into port-2 is directed to port-3 for detection. Circulators are passive devices and not vulnerable to external disturbances (except for very strong local magnetic fields). The circulator is probably the only item on our shopping list that is not obtainable in a form that covers the full frequency range that we desire. Over the years, I created a wide collection through direct purchase, second-hand purchase, and removal from old equipment. For most practical purposes I suggest purchasing three circulators to cover the ranges 1–2, 2–6, and 6–18 GHz, such as the broadband coaxial SMA circulators (ca 3 × 400 €) from UIY (Shenzhen, PRC). Other vendors carry items of similar specifications. Exchanging the circulators to cover different frequency bands is a minor inconvenience compared to the exchange of resonators and dewars required with single-frequency bridges. The exchange may be automated by building a switching unit with multiple electronic RF switches.

The amplifier must boost the weak reflection signal from the resonator to a level that is optimal for detection by the RF diode. Under-amplification leads to poor signal-to-noise ratios; over-amplification may result in destruction of the diode. The balancing act is taken care of by the software within some limitations imposed on the amplifier. Amplification should typically be of order 20 dB, and saturation (in RF language: the P_{1dBm} point) should preferably occur below the maximally allowed input power level of the diode. Single amplifiers with these specifications and covering a frequency range of 1–18 GHz or more are available at a price. Practically, I suggest purchasing two amplifiers, e.g., to cover the frequency ranges 0.5–8 and 6–18 GHz such as the coaxial broadband amplifiers (together ca 600 €) from Mini-Circuits (Brooklyn, NY, USA). Amplifiers are active devices and must be connected to a cooling sink.

The amplifier is followed by a SPDT (single pole double throw) RF switch for DC-18 GHz (ca 500 € new; ca 50 € from eBay) to change between dip tuning via a power meter and EPR recording via the spectrometer's signal channel. The RF switch is manually operated by means of a power supply (on-off).

The power meter is an NI USB-5681 (ca 7000 €) from National Instruments (Austin, TX, USA). It operates from 10 MHz to 18 GHz with input up to +20 dBm. It has a maximum damage level of +30 dBm, which should well exceed the maximum output level of the RF amplifier. The power meter is the most expensive item on our shopping list. Other vendors carry similar meters for comparable prices. In principle, the dip tuning could also be performed, in a much cheaper way, on the voltage output of the detection diode (below); however, each diode should then be calibrated (dBm; voltage out) at regular intervals with a power meter. Also, the power meter can be attached to the output of any RF component in the bridge, and thus may prove to be indispensable for identifying a malfunctioning component.

The purpose of the diode is to detect the microwave signal reflected from the resonator and then amplified, and to convert it into a 100 kHz video signal that can be fed into the signal channel for demodulation, and whose average amplitude can be monitored by the PC for frequency-dip tuning. I use a PE 8013 (ca 1000 €) from Pasternack Enterprises (Irvine, CA, USA), which is a zero-bias Schottky detector with flat response over its 10 MHz to 18.5 GHz frequency range. Also, its transfer function for voltage versus incident microwave power is linear over the power range of our interest (0–20 dBm) and quasi-linear at lower values (cf Figure S3 in [5]). Its response compares favorably with that of other diodes that I have tried, and its stability in my hands is also better than tunnel diode detectors. Other vendors may, however, carry items of similar specifications. Note that the diode may be destroyed by application of excessive power. The maximum working power for the PE 8013 is 20 dBm and the maximally allowed peak input is 27 dBm, which is why we chose an amplifier that saturates around 20 dBm.

All RF components can be connected with short pieces (10–20 cm) of male-male RG402 coax cable except for the connection between the circulator and the resonator cell inside the magnet, which may vary depending on the positioning of the broadband bridge, and, especially, on the employed frequency. The connection may be made as short as possible (say 30 cm) to minimize losses at the highest frequencies; it can be made longer to increase the number of frequency dips (see Section 3.4) at lower frequencies, up to a few meters around 1 GHz, and tens of meters towards 0.1 GHz (cf [4]). An overview picture of the actual components is given in Figure 3, which also specifies a few required adaptors/connectors. I purchased all these connection items via eBay mainly from vendors in the PRC at an overall cost below 100 €.

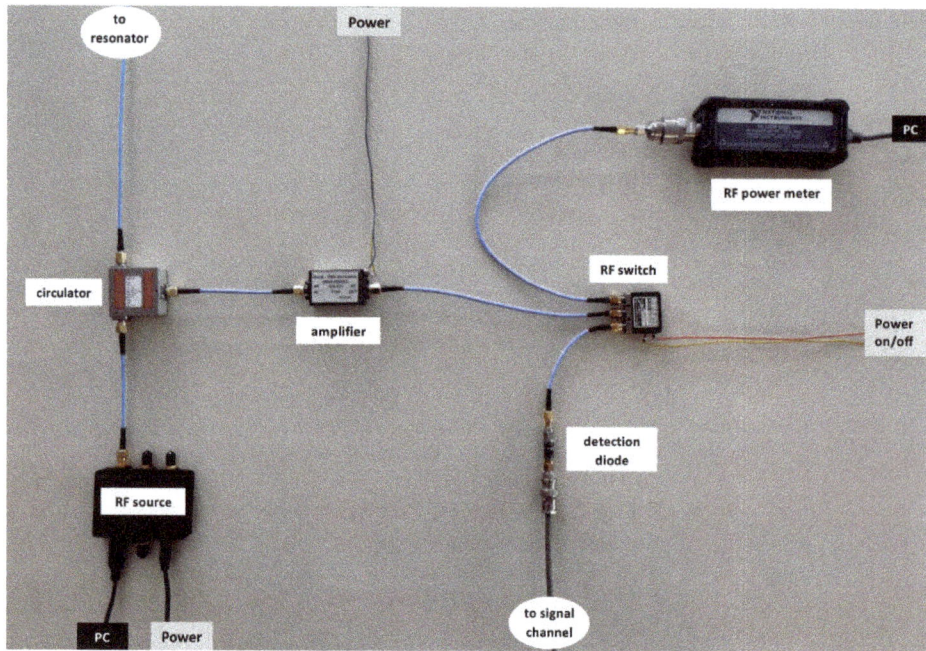

Figure 3. Photograph of the actual components of the broadband microwave bridge. The RF source and RF amplifier are active components and should be protected from overheating, e.g., by attaching to an aluminium plate (here omitted). 'Power' is a connection to a low-voltage power supply. 'PC' is a connection to a dedicated computer. The shortest coax cable, e.g., between circulator and amplifier, is 10 cm. Electrical grounding of hand wrists is strongly recommended before assembly.

3.3. Details of Home-Made Parts

A description of the resonator/sample holder has been given before and is reproduced here for convenience (Figure 4). The basic ingredients are always the same: a metal base plate of 1 mm thickness (not a critical dimension) is cut from a copper or aluminium plate. A female SMA connector is soldered to one end (the end of the aluminium cell is surrounded by a piece of 25 µm copper foil to allow soldering). The cell is insulated with 4–5 windings of 50 µm yellow acrylate tape of 5 cm width; this also fixes the copper foil onto the aluminium plate. Tape of other, especially darker, colors may cause background signals from, e.g., radicals and manganese. A rectangular cell compartment (sample compartment) is cut out of the tape, e.g., with a snap-off blade knife, down to the bare metal with dimension along the long cell axis that fits into the homogeneous field of the modulation coils. For doubling of the EPR signal amplitude, a mirror sample compartment may be cut out on the other side of the cell, whereby powder samples may later be held in place with, e.g., Parafilm wrapping. Then, a 0.2–0.25 mm wire of bare copper, or lacquered copper, or bare silver, is soldered to the inner conductor of the SMA connector and led over the cell (fixed in place with stripes of acrylic tape) to the sample compartment. It is then hand-wound around the sample compartment such that adjacent windings do not touch (separation space approximately equal to the wire diameter). From the end of the sample compartment, the wire is fixed with acrylic tape with a short lead towards the end of the cell. At no point are the wire and the base plate in electric contact. They form a microwave transfer line with full reflection at the open end. Note that for metalloproteins, aluminium is preferred over copper for the ground plate, and silver is preferred over copper for the wire, because aqueous protein solutions (that is, before the freezing act) have a significant tendency to liberate copper ions from solid copper, which in the spectroscopy results in Cu(II) background signals [5].

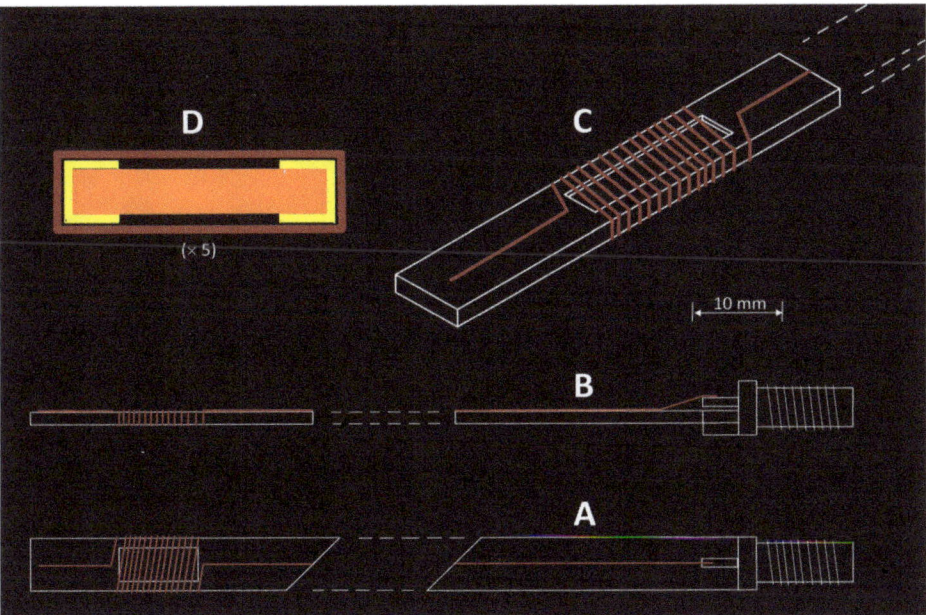

Figure 4. Isometric drawing of a wire micro strip cell. (**A**) Top view: (**B**) side view; (**C**) 3D view; (**D**) cross view. The cross view shows acrylic tape isolator (yellow) with sample compartments on both sides (black) surrounded by a low-permittivity wrap (brown), e.g., parafilm, or an extra layer of acrylic tape, to keep the sample in place. See the main text for further details. This figure has previously been given as Figure S1 in the Supporting Information to [4].

Figure 5 shows 2D drawings of resonators in three different forms. The simple, small cell (6.5 mm wide; sample compartment 4×16 mm^2) is intended for setup testing, calibrating, familiarizing purposes. Its dimensions are limited such that it can be placed inside a regular X-band cavity so that the original coils integrated in the side walls of the cavity can be used for field modulation, and no home-made modulation coil assembly is required. Note that in this approach, the cavity is simply a housing and is not used for microwave storage. The intermediate-size cell in Figure 5 is the standard research cell for low-temperature work. It fits into the flow dewar, which in turn fits into the home-made modulation-coils assembly. The wide cell in Figure 5 is to optimize the EPR signal from powders of maximized sample size at room temperature when no flow dewar is required. Note that these cells were not designed to accept unstable samples. Nitrogen cooling of cells (cf [4]) holding aqueous-solution samples in a glove box will provide protection against denaturation by air.

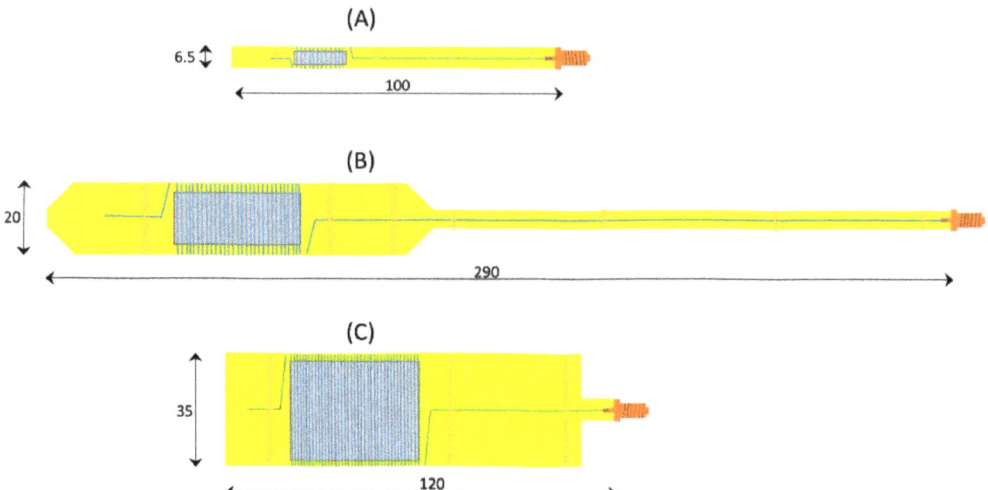

Figure 5. Geometries of the resonator cell intended for different applications. (**A**) Mini cell with small sample compartment and with dimensions that allow its placement in a rectangular X-band cavity, intended for learning and calibration purposes; (**B**) standard cell whose sample compartment fits within the homogeneous-field space of the dedicated modulation-amplitude coils, and whose dimensions allow it to be used in combination with a dedicated He/N$_2$ flow dewar; (**C**) oversized cell with maximized sample compartment for use at ambient temperature.

In the last described version of the broadband machine, I used a ready cylindrical modulation-coils assembly that was taken from a Varian E-line Q-band spectrometer. My choice was based on the fact that it fitted an existing helium-flow dewar for the Q-band setup, to which the dimensions of the low-temperature broadband resonator were adjusted [11]. Obviously, these items are not generally available, hence the need to construct similar structures in-house. A picture of the home-made modulation-coils assembly is in Figure 6. The coils were obtained as brandless mobile phone inductive charging units (20 W) whose regulatory printed circuit boards were disconnected and discarded. The coil wire (12 turns) is of the Litz type made of 100 enameled thin copper wires each of 0.09 mm diameter. The coils were wire-connected in series and mirror-image folded over a plastic (Teflon) cylinder, with an inner diameter of ca 40 mm, such that their individual fields will add. When you are sure you have the right orientation, the coils can be provisionally attached to the cylinder with transparent tape. Sturdiness is then provided by surrounding the coils with heat shrink tubing of proper size. The leads to the coil pair should be connected to a twin BNC to double wire jack (Figure 6), so that it can be connected to the twin BCN male plug of the spectrometer's modulation cable. The quality factor of the coil pair should be optimized for use with 100 kHz modulation. To this goal, measure the coils' impedance, L (in Henry), here: 17 µH, with an impedance meter, or a multimeter with impedance option, and use the following well-known tank circuit resonance equation to calculate a compensating capacitance, C (in Farad), for a modulation frequency ν (in Hertz), where typically ν = 100 kHz,

$$LC = \left(\frac{1}{2\pi\nu}\right)^2 \tag{6}$$

which should be added in parallel to the coils circuit in the form of a polypropylene film capacitor or a ceramic disc capacitor (do not use polarized capacitors). Then, fix the assembly in the gap of the magnet of the EPR spectrometer, e.g., with two pieces of polyurethane. For studies at ambient temperatures, the modulation cylinder may be placed horizontally inside the magnet; in combination with flow cryogenics, the orientation will generally be vertical. The spectrometer's internal signal channel calibration procedure can now be employed to calibrate the coils using a BDPA sample (see Section 3.6). For the coils in Figure 6, this gives a maximum modulation amplitude of 13.5 gauss at 100 kHz. Modulation-field homogeneity was tested with two cells of type B (Figure 5), one of which had a spot sample of BDPA in the center of the sample compartment, and the other had five spots of BDPA in an X-form over the sample compartment. Both cells gave a single-line spectrum with peak-to-peak width of 0.7 gauss at 462 MHz.

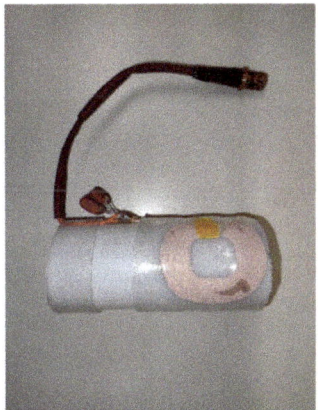 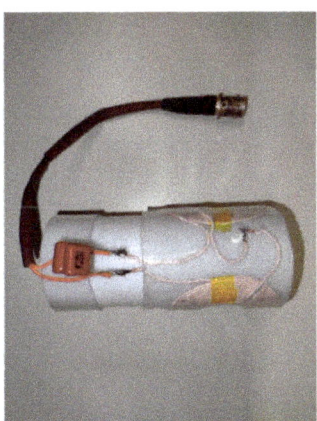 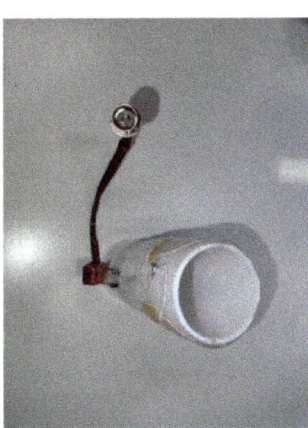

Figure 6. Photograph of modulation-coils assembly. The construct is based on a plastic cylinder on which two Litz-wire coils, taken from cell phone chargers and connected in series, are folded and attached. The diameter of the cylinder is such that it accommodates either room-temperature cell C of Figure 5 or a cooling system in combination with cell B of Figure 5. Note the film capacitors, with C = 100 + 68 nF, that tune the tank circuit to resonance at 100 kHz, and the twin BNC jack for connection to the spectrometer's modulation cable.

A simple flow dewar (colloquially known as 'the Swedish system') for Q-band is described in [11], which was designed for magnets on a stand accepting a 30–50 L helium vessel underneath (as used in our lab). This simple dewar is topped with rubber corks and a plastic (PVC) elongation tube in which the low-T cell of Figure 5 can have its sample compartment cooled by helium flow, while its thin top extension is subject to a heat gradient towards ambient temperature. This system was used in my previous broadband EPR studies [4–6]. Unfortunately, there is no universal solution for the adaptation to broadband EPR of the various flow cryostats in use in different laboratories. Perhaps the most commonly employed systems are those from Oxford Instruments (Abingdon, UK). Their CF935P cryostat for Q-band cavities has an inner diameter of 43 mm, which is more than enough to accept our standard low-temperature cell (Figure 5B), or even the oversized cell (Figure 5C). Its closing flange (KF50) would allow for the construct of a simple adapter to accommodate broadband cells. Modification of their ESR900 cryostat for X-band cavities would require a rather more challenging replacement of its quartz dewar by an elongated glass dewar with a wide top.

3.4. Tuning Procedure

A user interface for the dedicated PC, written in LabVIEW, is employed for tuning the broadband bridge. The program can be found in Supplementary Materials. A screenshot is given in Figure 7. After start-up, the program asks for the limits of a frequency range to be explored, which is then scanned at a low incident microwave power of −10 dBm, providing a pattern of potential frequency dips as exemplified in Figure 7. The user should then bracket any one of the dips, using a cursor, and re-scan (Figure 7). The chosen dip should preferably have its minimum below the input power. The frequency limits of the dip should then be refined to zoom in on a depth of ca 4 dB, and the input power should be raised such that the dip minimum corresponds to a power reading of ca 0–10 dBm (Figure 7). The source should now be set to radiate continuously at the frequency of the dip minimum and at the corresponding input power ('operate'). The RF switch can now be engaged to route the reflected microwave to the detection diode.

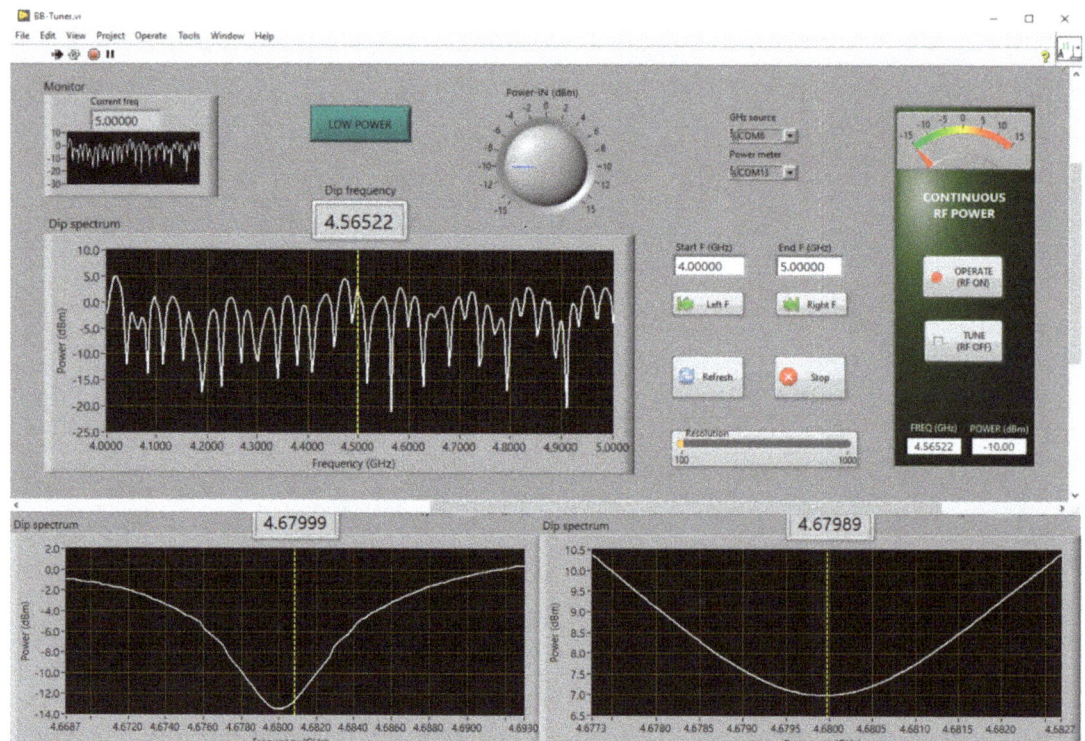

Figure 7. Screen shot of the dip-tuning program. The LabVIEW program monitors the signal from the power meter as a function of the scanned microwave frequency. (**Upper panel**) After an initial choice of frequency limits (here: 4–5 GHz), the program produces the power response at −10 dBm input power. A cursor is then used to single out an individual dip, using the 'left' and 'right' buttons; (**lower left panel**) this dip peak is re-scanned at low power and is then further reduced, using the cursor, to limiting power levels of circa 4 dB above the dip power; (**lower right panel**) input power is raised such that a dip power ensues that will be optimal for the detection diode, that is, circa 0–10 dBm. The bridge is now tuned and can be turned to permanent RF output ('operate'), and reflected power can be switched, with the SPDT switch, to the detection diode for EPR spectroscopy.

The spectrometer can now be run with its proprietary software in which the X-band bridge is not engaged (that is X-band tuning is at 'stand by'). The spectrum is stored in the standard way, e.g., the resulting file(s) are Bruker .DSC (=description in ascii) and .DTA (=data in binary) files. The files are identical in every respect to files generated with the X-band bridge engaged except that the microwave frequency value in the .DSC file is zero. To fix this, open the DSC file, e.g., with Windows Notepad, and replace 'MWFQ 0.0' with the actually used frequency in Hz in Bruker notation, e.g., MWFQ 9.123456e+09.

3.5. Small-Sample Testing: Mn^{2+} in CaO

Employing the straight, mini test resonator (see Figure 5A), the broadband spectrometer can be tested without the need to build a modulation coil assembly, that is, by using the X-band cavity resonator, not as a resonator, but only as a broadband-cell holder with modulation coils in its side walls. The purpose of this setup is to provide the operator with a convenient way to check the success of the spectrometer conversion and to become familiar with the use of the instrument. It can also be used for magnet field and scan calibration purposes. The setup is limited in its sensitivity by the limited sample size (16 × 4 × 0.15 mm^2) of the resonator, and it does not allow for cryogenic measurements. An illustrative example is given in Figure 8 using an Mn contamination in CaO. The commercial CaO sample has a quoted metal-based purity of 99.95%, therefore the Mn contamination is <0.05%. Manganese is 100% ^{55}Mn with electron spin $S = 5/2$ and a nuclear spin $I = 5/2$, which splits the main $m_S = |\pm 1/2\rangle$ line from substitutional Mn^{2+} in the cubic CaO lattice into a six-line pattern. The combination of a small line width (i.e., strong signal) and relatively low Mn concentration (i.e., weak signal) affords a system that poses a mild challenge in terms of attainable signal-to-noise ratio. The individual spectra of Figure 8 were obtained as 16 × 100 s scans except for the lowest two frequencies, which took 36 × 100 s. The figure is a clear illustration of the reduction in sensitivity that comes with decreasing the microwave frequency as a result of a reduced Boltzmann population difference over the electron spin levels.

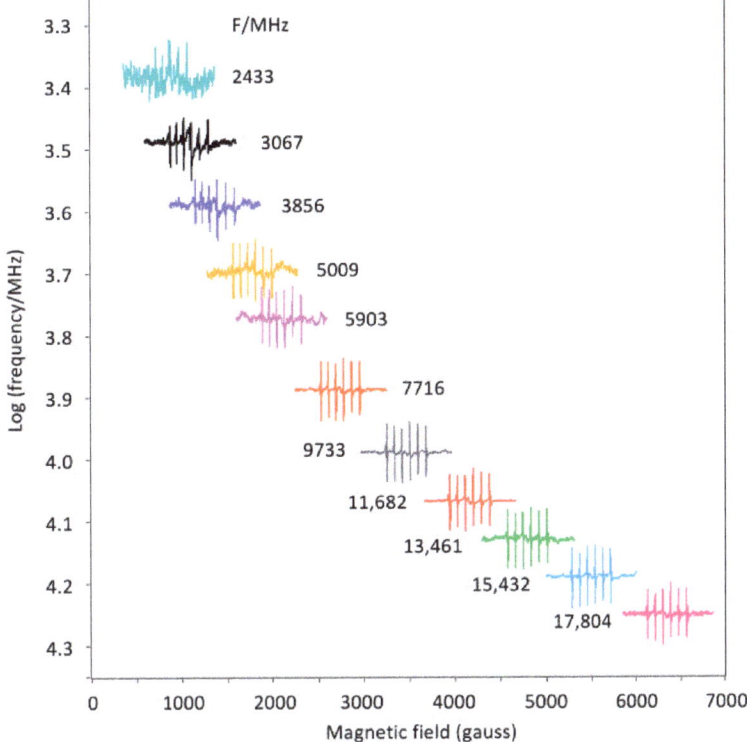

Figure 8. Broadband EPR of Mn^{2+} trace contaminant in cubic CaO. The room-temperature spectra were obtained at the indicated frequencies with the low-sensitivity mini cell A of Figure 5 and with three different circulators (2–6, 7–16, and 11–18 GHz). The 100 kHz modulation amplitude was 1 gauss. The figure illustrates the field range required for broadband EPR up to circa 18 GHz, and also the decrease of signal-to-noise ratio with decreasing frequency.

We now zoom in on the details of the spectrum taken at two significantly different frequencies of 17.8 and 3.9 GHz (Figure 9). The Mn^{2+} contaminant is substitutionally replacing Ca^{2+} in the cubic CaO lattice [12], so the spin Hamiltonian, with $S = 5/2$ and $I = 5/2$, is

$$\mathcal{H} = g\beta BS + ASI + \frac{1}{6}a[S_x^4 + S_y^4 + S_z^4 - \frac{1}{5}S(S+1)(3S^2 + 3S - 1)] \quad (7)$$

describing the electronic Zeeman interaction, the central hyperfine interaction, and the fourth-order cubic zero-field interaction [13]. Typically, a is much smaller than A [12], so the spectrum should consist of six hyperfine lines, whereby each line should be split into a symmetrical pentad from zero-field interaction. This latter fine structure is not observed (Figure 9), which presumably means that four of the five lines are broadened, due to a distribution in zero-field interaction strength. On the other hand, each central line ($|+1/2\rangle \longleftrightarrow |-1/2\rangle$) is unaffected by the magnitude of a. Observed single satellite lines (Figure 9) are symmetrically separated about the center ($\approx g$ value) of the spectrum, and they are presumably from a slightly different Mn^{2+} site in these microcrystals of small dimensions. The satellite lines are frequently observed, but not commented on, in 'powder' samples of Mn^{2+} in CaO (e.g., [14]).

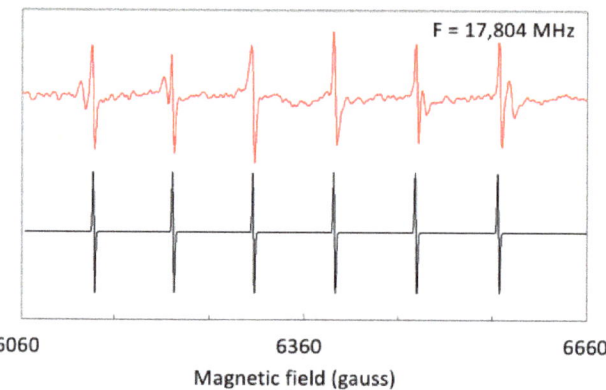

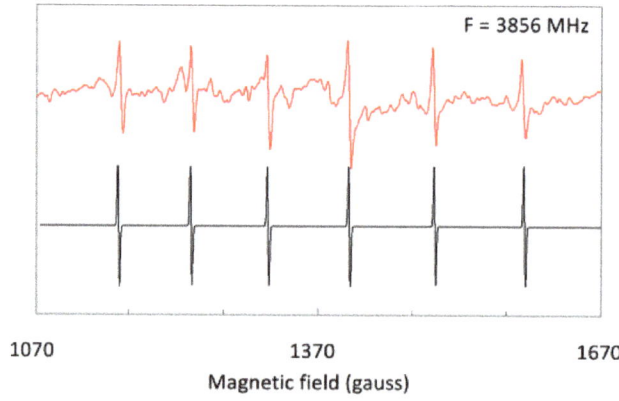

Figure 9. Comparison of hyperfine details from Mn^{2+} in CaO at 17.8 versus 3.9 GHz. The 100 kHz modulation amplitude was 1 gauss. The spectra at 17,804 and 3856 MHz are compared to identify frequency-dependent changes. With the help of simulations of the main signal from the transition within the $m_S = |\pm 1/2\rangle$ doublet, a deviation from field-equidistant hyperfine splittings is found, which increases with decreasing frequency.

Simulations of the transition within the $m_S = |\pm 1/2\rangle$ doublet (A = 85.7 gauss; a is undetermined) illustrate spectral-shape dependence on frequency. At 17.8 GHz, the hyperfine lines are almost equidistantly separated; the splittings at lowest and highest field differ by 5% only. In other words, the hyperfine interaction is close to a first-order perturbation of the Zeeman interaction. Contrarily, at 3.86 GHz the difference is increased to 21%. The hyperfine splitting pattern has clearly become asymmetric. The g value obtained from simulation is g = 2.0023 at 17.8 GHz, g = 2.0010 at 9.76 GHz (not shown), and g = 1.9968 at 3.86 GHz, and since there is no reason why the g value should be frequency-dependent, this suggests that the fields near $g \approx 2$ at the extreme frequencies require calibration correction (see below).

3.6. Field Calibration with DPPH and BDPA

Polycrystalline DPPH (2,2-diphenyl-1-picrylhydrazyl) is an indefinitely stable S = 1/2 radical with a single-line spectrum in the X-band (g = 2.0036) of circa 2–3 gauss peak-to-peak linewidth from exchange narrowing and therefore high signal-to-noise ratio, which has been widely and extensively used as a field marker and sensitivity marker. Its use in high-frequency EPR has been under debate where several authors report asymmetry or resolution of spectral structure at higher frequencies [15–18] while others find a single line of width proportional to the frequency [19]. Differences in g anisotropy may be related to different synthetic routes [20]. Due to its strong signal, DPPH is easily measurable with a mini test resonator in the present setup. Using a sample recently obtained from Sigma-Aldrich, I measured the spectrum at frequencies near the extremes of the available range (Figure 10 top). Here, broadband EPR proves its merit for deconvolution of frequency-independent and frequency-dependent interactions.

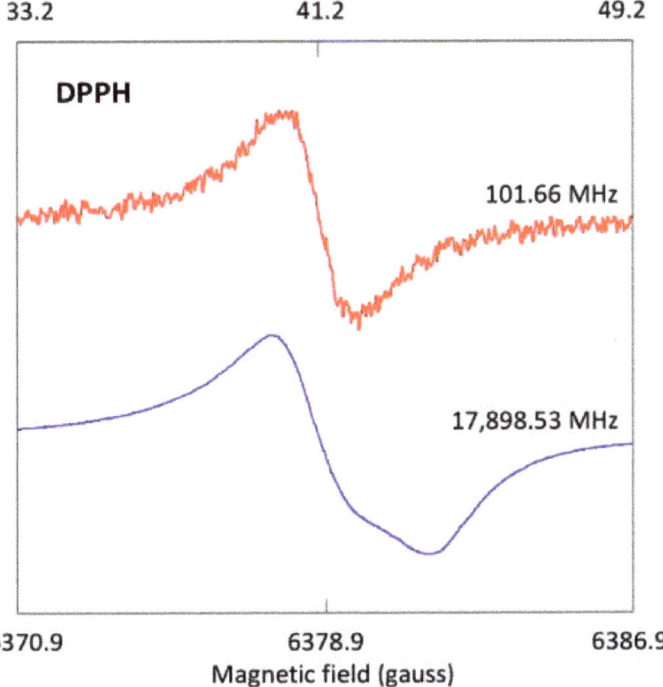

Figure 10. *Cont.*

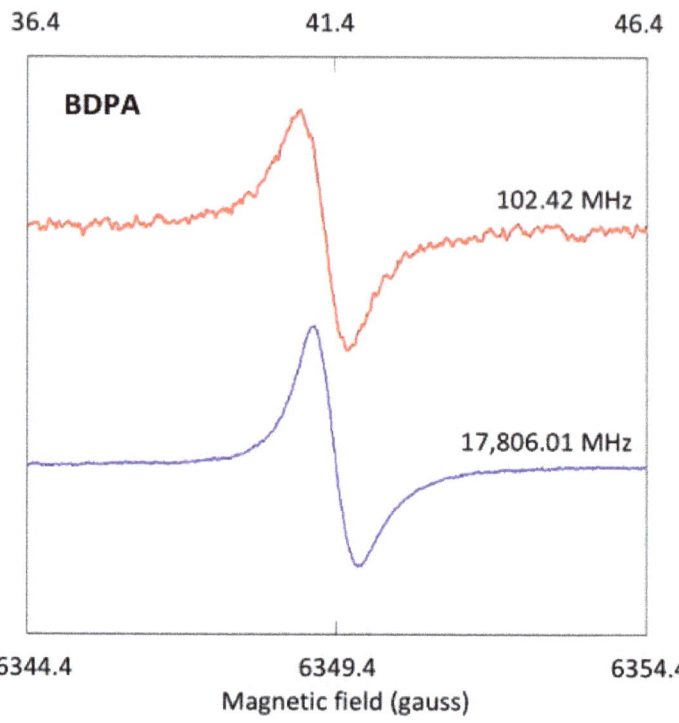

Figure 10. Frequency dependence of DPPH and BDPA calibration probes. DPPH (**top**) and BDPA (**bottom**) were measured in the mini cell of Figure 5A at two extreme microwave frequencies. The spectrum of DPPH can be seen to develop anisotropy (resolution of g matrix) at 17.9 GHz, while the spectrum of BDPA remains essentially isotropic over the full frequency range. BDPA is thus a suitable probe for accurate field calibration in broadband EPR, and DPPH is not.

At 102 MHz, the spectrum is a single line of 2.2 gauss width, consistent with earlier work down to a few MHz [21]. Contrarily, at 17.9 GHz, spectral anisotropy starts to resolve, which implies that its reliability as field marker will worsen with increasing frequency. Following field calibration at the X-band, the theoretical resonance field for $g = 2.0036$ at 102 MHz of 36.3 gauss is found experimentally to be 41.0, that is, off by +4.7 gauss. The corresponding field at 17.9 GHz is 6382.6 gauss, and the experimental field can only be estimated to be off by ca −3 to 4 gauss. Calibrations at higher frequencies will be increasingly uncertain. Recall that previously one high-frequency study of DPPH from the same manufacturer did not resolve anisotropy [19], so reproducible results over different batches are apparently not guaranteed.

A possible alternative for DPPH is the polycrystalline, stable $S = 1/2$ radical BDPA (α,γ-bisdiphenylene-β-phenylallyl) complex with benzene (1:1), which exhibits a single-line spectrum in the X-band ($g = 2.00254$) with even smaller peak-to-peak linewidth of circa 0.6–0.7 gauss, and which is recommended by EPR spectrometer manufacturer Bruker for modulation frequency calibration purposes [14]. For this compound, no anisotropy is resolved (Figure 10 bottom), which suggests that BDPA (at least from this batch) is an appropriate field calibration marker for broadband EPR over the available frequency range. The field deviations of the EMXplus 9 kgauss magnet, found with BDPA, are +4.7 gauss at 102 MHz and −3.6 gauss at 17.9 GHz.

4. Discussions

Over the last decade, I have developed a broadband continuous-wave EPR spectrometer for the approximate frequency range of 0.1–18 GHz. In the present paper, the building of the instrument has been simplified to the implementation of a conversion kit into existing X-band spectrometers. This leads to a considerable reduction in construction costs and complexity. The elements of the kit consist of readily purchasable parts plus a few items that have to be manufactured in-house. Fabrication of a simple mini cell is the only requirement for initial testing and getting acquainted with the setup. The complete kit is universally applicable, except for the adaptation of the cryogenics, whose geometry will depend on the targeted X-band spectrometer. Operation of the modified spectrometer requires only minor additional instruction.

In terms of application, the reader is advised that the broadband spectrometer is not intended to replace single-mode resonator spectrometers. Application of the described extension should rather be seen as a natural follow-up to initial studies with the standard X-band spectrometer. The latter is superior in sensitivity, although this may not necessarily be true for cavity spectrometers at other frequencies. Also, attainable energy density in the broadband resonator is less than in X-band cavities with their high-quality factors, which makes it more difficult to saturate signals. Thus, if power saturation characteristics are desired, they should be determined in the standard spectrometer. Also, since operation frequency and dip characteristics may depend on exact cell geometry and sample permittivity, quantitation (that is, spin counting) versus a standard compound of known concentration is more readily achieved with the standard spectrometer. The broadband extension is specifically intended for spectral analysis from systems with all but the simplest spin Hamiltonians. A broadband data set provides a much more rigorous test for interpretation of the details of spectral shapes and line-broadening mechanisms.

5. Materials and Methods

DPPH (2,2-diphenyl-1-picrylhydrazyl), and BDPA (α,γ-bisdiphenylene-β-phenylallyl) complex with benzene (1:1) were obtained from Sigma-Aldrich of Merck (Amsterdam, The Netherlands). Mn:CaO was obtained from Alfa Aesar of Fisher Scientific (Landsmeer, The Netherlands) as calcium oxide 99.95% (metals basis). Silver wire was from Alfa Aesar as soft, annealed silver wire, 0.25 mm in diameter, 99.9% (metals basis). The standard X-band spectrometer is a Bruker EMXplus (Bruker Physik AG, Karlsruhe, Germany). All hardware for its conversion into a broadband spectrometer is detailed in Section 3. Used software was written in LabVIEW (2020) and is given in Supplementary Materials. For proper operation, the BB-Tuner program requires (free) downloading from the NI site and installing of the driver for the NI USB-5681 power meter. The Z-Calculator, for resonator design, does not require additional software.

Supplementary Materials: The following supporting information can be downloaded at: https://www.mdpi.com/article/10.3390/molecules28135281/s1, a LabVIEW program to calculate the characteristic impedance Z_0 for a wire micro strip resonator, and a LabVIEW program to tune the broadband-adapted X-band spectrometer (requires LabVIEW 2020 to be installed).

Funding: This research received no external funding.

Data Availability Statement: The data are contained within the article.

Acknowledgments: I thank Edward Reijerse for insightful discussion on modifiability of different helium flow systems.

Conflicts of Interest: The author declares no conflict of interest.

Sample Availability: Samples of the compounds are not available from the author; they can be obtained from the cited commercial sources.

References

1. Krzystek, J.; Ozarowski, A.; Telser, J. Multi-frequency, high-field EPR as a powerful tool to accurately determine zer0-field splitting in high-spin transition metal coordination complexes. *Coord. Chem. Rev.* **2006**, *250*, 2308–2324. [CrossRef]
2. Möbius, K.; Savitsky, A. High-field/high-frequency EPR spectroscopy in protein research: Principles and examples. *Appl. Magn. Reson.* **2023**, *54*, 207–287. [CrossRef]
3. Hagen, W.R. Broadband transmission EPR spectroscopy. *PLoS ONE* **2013**, *8*, e0059874. [CrossRef] [PubMed]
4. Hagen, W.R. Broadband tunable electron paramagnetic resonance spectroscopy of dilute metal complexes. *J. Phys. Chem. A* **2019**, *123*, 6986–6995. [CrossRef] [PubMed]
5. Hagen, W.R. Very low-frequency broadband electron paramagnetic resonance spectroscopy of metalloproteins. *J. Phys. Chem. A* **2021**, *125*, 3208–3218. [CrossRef] [PubMed]
6. Hagen, W.R. Low-frequency EPR of ferrimyoglobin fluoride and ferrimyoglobin cyanide: A case study on the applicability of broadband analysis to high-spin hemoproteins and to HALS hemoproteins. *J. Biol. Inorg. Chem.* **2022**, *27*, 497–507. [CrossRef] [PubMed]
7. Bagguley, D.M.S.; Griffiths, J.H.E. Paramagnetic resonance and magnetic energy levels in chrome alum. *Nature* **1947**, *160*, 532–533. [CrossRef]
8. Jiang, J.H.; Wu, D.L. Ice and water permittivities for millimeter and sub-millimeter remote sensing applications. *Atmos. Sci. Lett.* **2004**, *5*, 146–151. [CrossRef]
9. Napijalo, M.L.J.; Nikolić, Z.; Dojčilović, J.; Napijalo, M.M.; Novaković, L. Temperature dependence of electric permittivity of linear dielactrics with ionic and polar covalent bonds. *J. Phys. Chem. Solids* **1998**, *59*, 1255–1258. [CrossRef]
10. Xie, C.; Organov, A.R.; Dong, D.; Liu, N.; Li, D.; Debela, T.T. Rational design of inorganic dielectric materials with expected permittivity. *Sci. Rep.* **2015**, *5*, 16769. [CrossRef]
11. Albracht, S.P.J. A low-cost cooling device for EPR measurements at 35 GHz down to 4.8 ^{0}K. *J. Magn. Reson.* **1974**, *13*, 299–303.
12. Shuskus, A.J. Electron spin resonance of Fe^{3+} and Mn^{2+} in single crystals of CaO. *Phys. Rev.* **1962**, *127*, 1529–1539. [CrossRef]
13. Low, W. Paramagnetic resonance spectrum of manganese in cubic MgO and CaF_2. *Phys. Rev.* **1957**, *105*, 793–800. [CrossRef]
14. Weber, R.T. *Xenon User's Guide, Version 1.3*; Bruker BioSpin Corporation: Billerica, MA, USA, 2011.
15. Hutchison, C.A.; Pastor, R.C.; Kowalsky, A.G. Paramagnetic resonance absorption in organic free radicals. Fine structure. *J. Chem. Phys.* **1952**, *20*, 534–535. [CrossRef]
16. Kikuchi, C.; Cohen, V.W. Paramagnetic resonance absorption of carbazyl and hydrazyl. *Phys. Rev.* **1954**, *93*, 394–399. [CrossRef]
17. Lynch, W.B.; Earle, K.A.; Freed, J.H. 1-mm wave ESR spectrometer. *Rev. Sci. Instrum.* **1988**, *59*, 1345–1351. [CrossRef]
18. Earle, K.A.; Tipikin, D.S.; Freed, J.H. Far-infrared electron-paramagnetic-resonance spectrometer utilizing a quasioptical reflection bridge. *Rev. Sci. Instrum.* **1996**, *67*, 2502–2513. [CrossRef]
19. Krzystek, J.; Sienkiewicz, A.; Pardi, L.; Brunel, L.C. DPPH as a standard for high-field EPR. *J. Magn. Reson.* **1997**, *125*, 207–211. [CrossRef] [PubMed]
20. Žilić, D.; Pajić, D.; Jurić, M.; Molčanov, K.; Rakvin, B.; Planinić, P. Single crystals of DPPH grown from diethyl ether and carbon disulphide solutions—Crystal structures, IR, EPR and magnetization studies. *J. Magn. Reson.* **2010**, *207*, 34–41. [CrossRef] [PubMed]
21. Garstens, M.A.; Singer, L.S.; Ryan, A.H. Magnetic resonance absorption of diphenyl-picryl hydrazyl at low magnetic fields. *Phys. Rev.* **1954**, *96*, 53–56. [CrossRef]

Disclaimer/Publisher's Note: The statements, opinions and data contained in all publications are solely those of the individual author(s) and contributor(s) and not of MDPI and/or the editor(s). MDPI and/or the editor(s) disclaim responsibility for any injury to people or property resulting from any ideas, methods, instructions or products referred to in the content.

Article

A Review of the Dawn of Benchtop EPR Spectrometers—Innovation That Shaped the Future of This Technology

Vasily V. Ptushenko [1,2,*] and Vladimir N. Linev [3]

1. A.N. Belozersky Institute of Physico-Chemical Biology, Lomonosov Moscow State University, 119992 Moscow, Russia
2. N.M. Emanuel Institute of Biochemical Physics, Russian Academy of Sciences, 119334 Moscow, Russia
3. LINEV Group, LINEV Systems UK Limited, London SW1Y 5JG, UK
* Correspondence: ptush@belozersky.msu.ru

Abstract: By the early 1980s, unique devices appeared in the USSR: a series of benchtop specialized EPR spectrometers. This equipment was quickly accepted not only in science but also in medicine and in many technical and economic areas including chemical industries and geologic exploration. The appearance of these devices was perceived as a salvation for the Soviet magnetic resonance (MR) scientific instrumentation by those who worked in the field of EPR spectroscopy in the USSR. (However, the program of MR scientific instrumentation ceased to exist along with the USSR a few years later). The Belarusian State University in Minsk was the center of these developments. At that moment and for many years afterwards, these devices were unique with no analogues in the worldwide EPR industry. They remained the only mass-produced MR spectrometers on the territory of the former USSR after its collapse. For the first time, based on archival materials, patents, and our personal memoirs, we describe the development of these EPR spectrometers and discuss the most original technical solutions and the scientific tasks solved with this equipment We also remember the participants of the work, showing the historical context of these events.

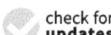

Citation: Ptushenko, V.V.; Linev, V.N. A Review of the Dawn of Benchtop EPR Spectrometers—Innovation That Shaped the Future of This Technology. *Molecules* **2022**, *27*, 5996. https://doi.org/10.3390/molecules27185996

Academic Editors: Yordanka Karakirova and Nicola D. Yordanov

Received: 15 August 2022
Accepted: 8 September 2022
Published: 14 September 2022

Publisher's Note: MDPI stays neutral with regard to jurisdictional claims in published maps and institutional affiliations.

Copyright: © 2022 by the authors. Licensee MDPI, Basel, Switzerland. This article is an open access article distributed under the terms and conditions of the Creative Commons Attribution (CC BY) license (https:// creativecommons.org/licenses/by/ 4.0/).

Keywords: EPR instrumentation; geologic exploration; chemical industry applications; EPR dosimetry; Belarusian State University

1. Introduction

In this paper, we describe the history of the creation of a series of scientific instruments, which is work that involved many people and research teams and played a significant role in scientific research in the field of EPR spectroscopy in the late 1970s–1980s and contributed to the USSR economy. However, the work involved in the creation of innovative scientific instruments, whilst essential for progress in science, often remains unnoticed by the scientific community, which is aware of only the final result. Still, a "master's tool" has been a subject of interest in all ages. We believe that this applies to the work and innovation of the EPR spectroscopy instruments.

This particular story has never been described in detail before; only a few early episodes have been reflected on to some extent in memories and discussions. In addition, despite the vast amount of work involved and the fact that a significant number of patents were obtained, virtually no articles were published. Therefore, to fill this gap, in this article, we describe the work involved based mainly on archival materials, patents, and personal memoirs of one of the authors. As this is the first description ever published, we dare to go into some historical details, focusing not only on the scientific tasks being solved but also on the technical details of the devices being developed as well as on the people participating in this work. We believe that the published memoirs of J. Hyde [1], who did not consider it shameful to include aspects of live communication between colleagues while telling about the technical history of "Varian", serve as a good example of how this mix of factual and personal aspects works well.

The history of the development of EPR spectrometers in the USSR described here is not an isolated one "without kith or kin". It is inscribed in the context of many years of research, development and instrumentation in the field, which began with the discovery of the EPR by Eugene K. Zavoisky in 1944 [2,3]. Some previous stages of the development of EPR spectrometers in the USSR were described in our earlier publications [4–7]. In this paper, we describe the last stage in the USSR experimental design program in the field of EPR, which later found continuation in the post-Soviet history of EPR instrumentation. The development appeared on the background of significant stagnation in the development and serial production of research-grade EPR spectrometers in the country which had become obvious by the mid-1970s.

The leaders in the field were Varian (USA) and Bruker (Germany), who developed and produced a set of unique EPR spectrometers which became state of the art standard for EPR spectroscopy. It was impossible to achieve the same level of instrumentation in the USSR with the outdated electronic components then available. Therefore, the salvation of Soviet EPR spectroscopy capability in these conditions required an innovative and creative approach rather than conventional design and technical solutions.

It is worth noting that although the article addresses the history of applied research, it remains relevant to the issues of modern economic theory of high-tech innovations, especially Christensen's Theory of Disruptive Innovation. The Department of Nuclear Physics of the Belarusian State University, where the development of benchtop EPR spectrometers started, was decades ahead of its time in terms of its scientific and institutional concept to innovation and development processes. In fact, it was an innovative incubator, i.e., a generator of high-tech ideas, which then resulted in establishing scientific laboratories, research institutes, and design centers.

The benefits of this scientific and economic model are confirmed, in our opinion, by the emergence of a group of high-tech companies including the LINEV Group, the producer of specialized benchtop EPR spectrometers, which considers itself the heir to the scientific and innovative traditions of the Department of Nuclear Physics.

2. Historical Background for the EPR Instrumentation in Minsk

In the 1950s, the main centers of EPR spectroscopy in the USSR were the Kazan Physical-Technical Institute (PTI; now E.K. Zavoisky PTI), the Moscow P.N. Lebedev Physical Institute (LPI), and the Moscow Institute of Chemical Physics (ICP; now N.N. Semenov ICP) of the Soviet Academy of Sciences (AS). The roots of the EPR instrumentation in Belarus were connected with these scientific centers. In 1956, the famous physicist Mikhail A. Eliashevich (1908–1996) moved to Minsk from the ICP in Moscow. In the ICP, chemical and biological EPR spectroscopy had been being developed since 1954, and Eliashevich was in the middle of events and actively participated in them [7]. In 1956, Stanislav Stanislavovich Shushkevich became his postgraduate student at the Institute of Physics and Mathematics of the Academy of Sciences of the Belarusian Soviet Socialist Republic (BSSR) (Figure 1). (Much later, in the 1990s, he became known as the head of Belarus who signed the Belovezh Accords with M.S. Gorbachev declaring that the USSR had effectively ceased to exist).

Eliashevich set his new laboratory team the task of experimentally mastering the detection of electron and nuclear magnetic resonance (EPR and NMR) and nuclear quadrupole resonance (NQR) spectra. To solve this problem, in early 1957, postgraduates S.S. Shushkevich and A.K. Potapovich went on an internship sat the LPI (specifically at the laboratory of A.M. Prokhorov, the future Nobel laureate) and ICP where the problem had been already had solved ([8], pp. 56–58).

In 1952–1953, Prokhorov, on behalf of the LPI director D.V. Skobeltsyn, switched to work that eventually led to the appearance of EPR masers [9]. For this purpose, A.A. Manenkov was transferred to LPI from Kazan PTI. In Kazan, A.A. Manenkov was a graduate student of B.M. Kozyrev, a colleague of E.K. Zavoisky [10], the discoverer of EPR. Thus, the EPR arrived in LPI from the "original source". At the same time, in the ICP, V.V. Voevodsky and L.A. Blumenfeld with their colleagues were developing EPR spectrometers suitable for chemical and biological works. The communication of Shushkevich and Potapovich with these scientific groups and the study of their experience made it possible to create in Eliashevich's laboratory in Minsk in the next three years a superheterodyne ESR spectrometer with two five-ton electromagnets made for this at the Leningrad "Electrosila" factory ([8], pp. 58–59).

Figure 1. Mikhail A. Eliashevich (**right**) with his disciple Stanislav S. Shushkevich (**left**) near the microwave assembly of their handmade EPR spectrometer. Minsk, 1957. Source: personal archive of V.N. Linev.

3. Development of Benchtop EPR Spectrometers: The First Steps

Having experienced the influence of key figures in EPR spectroscopy in late 1950s, the development of EPR instrumentation in Belarus bloomed and led to the concept (as well as the "incarnation") of small-sized EPR spectrometers a decade and a half later, when Shushkevich was already working at the Belarusian State University (BSU) as the head of the Department of Nuclear Physics and Peaceful Uses of Atomic Energy. His department, including the teaching staff and research laboratories, was the largest at the Physics Faculty of BSU (about 120 people) and, during its life, gave birth to several institutes and Special Design Departments (SDD). It specialized in applied instrumentation in the field of nuclear physics. Despite the obvious significance of this work today, its beginning was prompted by a rather accidental circumstance, namely, the attempts of the neighboring Department of Semiconductor Physics to develop a small-sized device for educational purposes. These attempts were based on the dead-end (as it turned out later) idea of using the pole tips of an electromagnet as the walls of a resonator. Actually, this approach did not allow obtaining a high Q-factor and hence high enough sensitivity of the resonator. V.F. Stelmakh, the head of the Department of Semiconductor Physics, asked Shushkevich to help with this work. The actual work was started in 1975 by one of the authors of this article, V.N. Linev, who had become a department member shortly before then. This task became the topic of his PhD thesis and later of his doctoral dissertation. (In the USSR, there was a two-stage system of academic degrees.)

By 1978, manufacturing of the first device was completed. It was housed in an oscilloscope case of ca. $274 \times 206 \times 440$ mm^3 size. Compare it with RE1306 EPR spectrometer—the main mass-produced EPR spectrometer in the USSR at that time—which consisted of four blocks. One of them, the electromagnet, was $1290 \times 1255 \times 710$ mm^3 in size; the other three blocks were slightly smaller but comparable in size [11]. Hence, the new equipment drastically surpassed existing models in compactness; however, its capabilities were limited.

It was obvious that to create a full-fledged scientific device rather than one serving for educational purposes only, significant efforts were still required. However, such a work demanded funding. Furthermore, before starting the development, it was necessary to assess the need for benchtop EPR spectrometers of Soviet science and industry. S.S. Shushkevich had a very wide network of acquaintances in Soviet radio spectroscopy, and he helped Linev to get in contact with some specialists in EPR spectroscopy who became the key customers and investors to drive the whole project. In Moscow, Linev met L.V. Bershov from the Institute of Geology of Ore Deposits, Petrography, Mineralogy and Geochemistry (IGEM) of the Academy of Sciences of the USSR and B.M. Moiseev from the All-Union Scientific-Research Institute of Mineral Resources (VIMS) of the Ministry of Geology of the USSR.

These institutes were engaged in the exploration of uranium deposits [12,13]. Leonid V. Bershov studied paramagnetic centers in minerals; he showed that they could be used in petrographic research to study the diastrophic blocks, e.g., determining their radiation history and conditions of their formation as a whole (see, e.g., [14]). Boris M. Moiseev studied paramagnetic Al- and E-centers in quartz crystals and determined the rate of their formation due to the radiation of different types resulting from the decay of uranium, thorium, potassium and other radioactive nuclei contained in the rock. This made paramagnetic properties of rocks an indicator for exploration of uranium deposits [15,16]. Thus, a new direction of paleodosimetry, a paleodosimetric method of searching for uranium deposits, was opened in geology. The detection of radioactive uranium ore lying at considerable depths by studying paramagnetic centers in the rocks at the surface layers thus became possible, and EPR allowed the delimiting of uranium deposits. However, the outdated models of EPR spectrometers adjusted for laboratory work, e.g., RE1301 [14] or RA-100 [15], remained the main instruments of these studies. The development of small-sized portable EPR spectrometers suitable for work in expeditions would significantly reduce the cost of and speed the exploration of minerals. Finally, L.V. Bershov and B.M. Moiseev found funding for the development of small-sized EPR spectrometers.

Several contracts were agreed between BSU and VIMS on the basis of long-term cooperation for at least 10 years in the future. An industrial scientific-research laboratory for magnetic resonance and gamma resonance spectroscopy of geologic materials was established in 1980 to carry out the development of small-sized EPR spectrometers. In 1985, V.N. Linev was appointed head of the laboratory [17]. Formally, since ca. 1977 until 1991, the laboratory developed and manufactured EPR equipment (e.g., [18]), as well as a wide range of other scientific instruments, for the Ministry of Geology of the USSR (Mingeo). Thus, for the first time ever, portable automated Mössbauer spectrometers (used for analysis of cassiterite, the main ore mineral for tin production, in ores) were developed, and the experimental factory of BSU allowed the manufacturing of a series of these devices.

We have covered these organizational issues to show the motives and method of organizing such a study in the USSR where direct commercial activity was impossible. Actually, the main challenges were related to innovating new technical ideas that were essential to achieve compact spectrometers. To some extent, the developers learned the technique of EPR instrumentation by studying the design of Varian and Bruker spectrometers available in the USSR. Copying these devices was out of the question due to the lack of necessary modern electronic components in the USSR coupled with the realization at an early stage by V.N. Linev that the simple miniaturization of the full-size spectrometers developed earlier was a dead-end. Compact devices had to be developed "from scratch" with essentially new solutions for each component of the instrument. Surprisingly, this project benefited from the fact that its participants were "amateurs" in the field of EPR spectroscopy. They were not "blinded" by conventional technical solutions, and this meant that they were able to innovate new solutions to the technical challenges.

4. Technical Solutions for Compactization of EPR Spectrometers

By far, the most massive element of a full-size EPR spectrometer is the electromagnet. In the most common Soviet X-band EPR spectrometer of that time, RE1306, the electromagnet weighed more than a ton, and for the next model RE1307, it was ca. 3.5 tons. For a portable instrument, it should not exceed 40–50 kg. The mass of the electromagnet grows rapidly with the increase in the size of the pole gap in which a uniform magnetic field must be provided. Hence, reducing the electromagnet mass by almost two orders of magnitude should be the first step toward creating a benchtop device, and this reduction depended on the minimum possible pole gap width. We mentioned above the attempt to use the pole tips of the electromagnet as the walls of the resonator. This challenging idea (called "EPR sensor") allowed a pole gap as narrow as 8 mm, but this approach turned out to be a dead end. The requirement to achieve a sufficiently high resonator Q-factor was incompatible with a monolithic "magnet-plus-resonator" design and demanded at least a 14 mm pole gap. According to this minimum distance and the requirements for maximum magnetic field inhomogeneity ($\leq 10^{-5}$), various systems to create and customize a magnetic field were developed.

Further technical findings involved using flat rectangular plates for magnetic field modulation and a modular resonator [19]; the former provided reducing the size of the resonator without reducing its Q-factor, while the latter allowed adapting the resonator to different tasks. This extraordinary design required taking into account the distribution of electric and magnetic fields along the assumed cutting lines of the resonator. The resonator was combined with a synchronous detector of original design (Figure 2). The replacement of the electromagnets conventionally used for EPR spectroscopy by permanent magnets-based systems was probably the most paradoxical technical solution (Figure 3). This solution allowed a reduction in the mass of the magnet by several orders (down to 8 kg!) and the energy consumption by an order of magnitude.

(a) (b)

Figure 2. EPR resonator combined with the synchronous detector, dismantled (**a**) and assembled (**b**). Source: Personal archive of V.N. Linev.

To regulate and sweep magnetic field *H* in the pole gap, various methods of mechanical and electro-mechanical *H* regulation were invented, employing moving the additional permanent magnet relative to the magnetic circuit, introducing additional adjustable gaps into the magnetic circuit, etc. [20–23] (Figure 4). This branch of experimental design was supervised by V.V. Lisovsky, a member of the BSU Department of Nuclear Physics. The electronic computing facilities available at the BSU at that time were used for mathematical simulation that preceded the experimental work. Along with permanent magnets-based EPR spectrometers, the electromagnet-based lineup was developed. The design of a compact electromagnet providing highly homogeneous magnetic field comprised modeling of the magnetic circuit and the pole tips configuration as well as the search for a material with optimal characteristics.

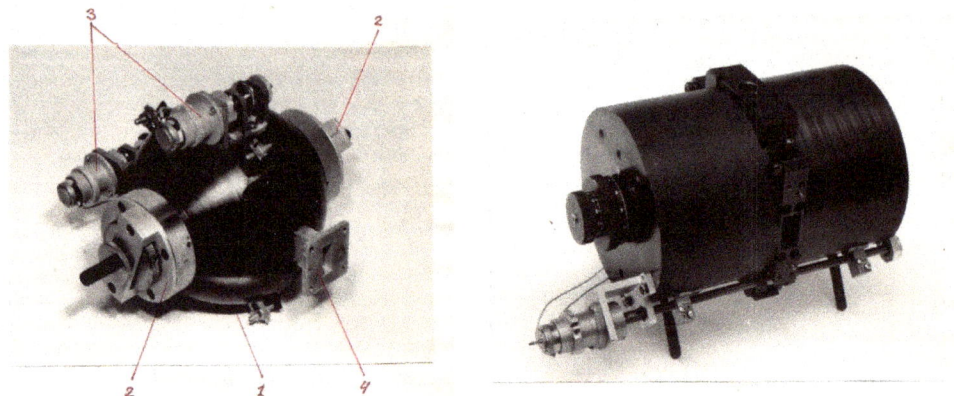

Figure 3. Two models of permanent magnet-based magnet blocks for benchtop EPR spectrometers. A piece of waveguide of X-band EPR spectrometer in the left photo indicates the miniature size of the magnet. The diameter of cylindrical magnet block in the right photo is 14 cm. *1*—magnet yoke, *2*—magnetic field adjustment mechanism, *3*—frequency tuning and coupling nodes of the measuring resonator, *4*—lead-in waveguide. Source: personal archive of V.N. Linev.

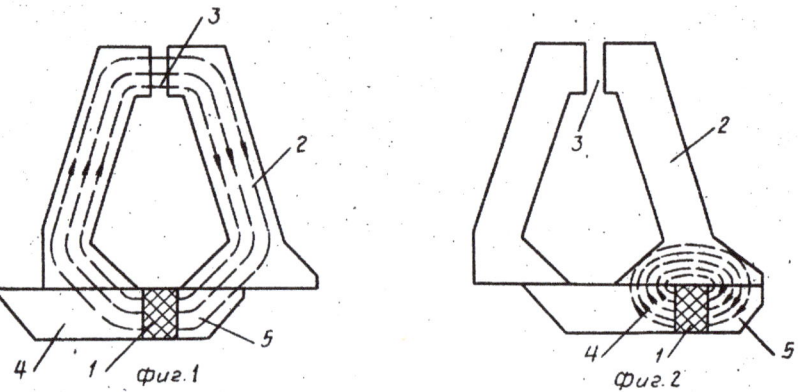

Figure 4. The scheme of mechanical regulation of magnetic field in the pole gap employing moving the additional permanent magnet relative to magnetic circuit [21]. *1*—permanent magnet, *2*—magnetic circuit, *3*—pole gap, *4,5*—movable part of magnetic circuit. This scheme provided a change in the magnetic field in the pole gap in the range 0.01–0.7 T.

The other massive component of the EPR spectrometer, a power source, was reduced significantly in size due to a reduction in the energy levels required. In a conventional EPR spectrometer, the electromagnet and microwave source (klystron) are among the main energy consumers. As the energy demand of the magnet was reduced, the klystron energy requirements became the limiting factor. To solve the issue, the time-tested klystron (which bore all the disadvantages of electro-vacuum devices) was replaced by a Gunn diode, which was a solid-state electronic element invented a decade earlier. The original Gunn diode-based X-band microwave generator, highly stable and tunable in frequency, was probably one of the key elements invented already at the start of the project and subsequently being improved over the following years. The second approach was to use a p-i-n diodes-based microwave attenuator. These inventions allowed reducing the total mass and energy demand of the spectrometer as well as improving the microwave durability.

To simplify the use of the device, automatic tuning systems were developed for all the units. One of these was an approach later called the Method of Adaptive Registration of EPR spectra. This involved sweeping the magnetic field with a varying speed depending on the spectrum slope [24]. This allowed skipping "empty" regions (free of EPR signals) during H sweeping without manual tuning of recording parameters and hence significantly increased the efficiency of recording multicomponent spectra (especially with long-term signal accumulation). The techniques for stabilization of the magnetic field and resonance conditions during long-term continuous measurements were developed as well as for on-the-fly signal accumulation and processing, which was necessary for EPR recording in the field [20,25]. All these solutions, in different combinations depending on the specific requirements and tasks, were embodied in a whole series of small-sized EPR spectrometers developed in the BSU Industrial scientific-research laboratory for magnetic resonance and gamma resonance spectroscopy of geologic materials in subsequent years.

Below, we describe the scope of application of small-sized EPR spectrometers and how it expanded (Section 5). However, each class of applications required the development of special technical solutions. As an example, a special resonator for magnetic semiconductor materials with magnetic bubble domains, a subject of great interest for the electronics industry, was developed (in fact, it was a ferromagnetic resonance (FMR) spectrometer). The size of the samples (usually large plates or films) did not allow them to be placed in the resonator. An alternative design of measurement techniques was developed instead: the sample was placed outside the resonator, and a coupling element between the sample and the resonator was inserted inside the latter. The sample holders provided precise movement of the sample relative to the coupling hole, which allowed obtaining a profile of magnetic defects in the sample. Thus, an FMR scanner for magnetic films was developed.

For biological samples, a special model of the resonator was also developed. It included a thick-walled ceramic or quartz tube of large diameter, which allowed partially shielding of the electric component of the field in the resonator. Resonators with one-, two- and three-axis goniometers were constructed for studying paramagnetic centers in crystals in IGEM. The built-in goniometers allowed precise positioning of the crystals inside the resonator. In addition, there was a method for measuring the anisotropy parameters of paramagnetic samples employing an alternating modulating magnetic field perpendicular to the polarizing magnetic field [26].

Eventually, a complex of original technical solutions for adjusting each of the nodes of the EPR spectrometer design for specific applications was developed [27,28]. The experience gained actually turned the BSU Industrial laboratory into one of the national-scale centers for the development and production of EPR spectroscopy equipment (Figure 5).

This detailed description of specific technical solutions is interesting in understanding the development of technology as a whole. Here, we can compare some different approaches in the drift of technical ideas. The need for benchtop EPR spectrometers was quite obvious. Two other groups of developers from the V.I. Ulyanov (Lenin) Leningrad Electrotechnical Institute (LETI) and from the Leningrad Association of Electronic Instrumentation "Svetlana" were solving similar challenges, i.e., the development of small-sized EPR spectrometers for oil production and refining. However, they started from "classic" EPR spectrometer design and moved toward its miniaturization. Probably, this approach based on the "scaling" of full-sized devices had more severe limitations. Thus, the minimum weight of the EPR spectrometer with this approach was about 75 kg (cf. 28 kg achieved in BSU) ([29], pp. 11, 20).

Figure 5. Among the participants of the XXIV Congress AMPERE on magnetic resonance and related phenomena in Poznan, 1988: J. Hyde (4th from the left), Yu.V. Yablokov (in the center), S.S. Shushkevich (4th from the right), V.N. Linev (3th from the right). Source: personal archive of V.N. Linev.

5. New Applications of Benchtop EPR Spectrometers

As the development progressed, the contacts of developers with potential users expanded, and positive feedback led to new applications of small-sized EPR spectrometers. One of the most fruitful came from the contact of V.N. Linev with Peter M. Solozhenkin, who was one of the leaders of the Academy of Sciences of the Tajik SSR. Tajik industry needed an instrument to control ore cleaning at metallurgical plants, and in 1978–1979, BSU produced a benchtop EPR spectrometer for this purpose. This work served as a catalyst for expanding the scope of benchtop EPR spectrometers. The developers started collaboration with Yuri A. Zolotov and his group from the V.I. Vernadsky Institute of Geochemistry and Analytical Chemistry (GEOCHI) of the Soviet Academy of Sciences. The interests of GEOCHI included the chemistry and geochemistry of uranium [30]. Although Yu. A. Zolotov had already collaborated with another EPR spectrometers manufacturer, the Leningrad Electrotechnical Institute [31], the new devices from BSU proved to be in demand. A while later, they proved to be in demand in Zelenograd, the Soviet "Silicon Valley", where FMR could be used to control the quality of materials with magnetic bubble domains (we described the development of an FMR scanner in the previous section). In Zelenograd, Anatoly Yu. Kozhukhar coordinated this collaboration [32,33].

An important direction in the EPR equipment development appeared due to BSU collaboration with Oleg P. Revokatov, Yuri M. Petrusevich, and their coworkers from Moscow State University. O.P. Revokatov suggested an idea to detect the cancer risk using spin probes in a blood test. Previously, Revokatov tried to employ NMR data on proton spin-lattice relaxation in human blood serum as a cancer diagnostic method [34], and he participated in the development of NMR equipment for the diagnostic testing of biological liquids. The new idea was based on employing the rotational mobility of spin probes (which is reflected in the EPR spectra) for assessing the formation of permolecular structures in blood.

All this work assumed clinical usage and hence required, in addition to compactness, easy-to-operate (i.e., automated) devices. Together with Shushkevich, Revokatov managed to interest the Belarusian Research Institute of Oncology and Medical Radiology (later awarded the name of N.N. Alexandrov and transformed to National Cancer Centre of Belarus") and, personally, Alexander A. Mashevsky and Violetta V. Prokhorova, in this idea. After the Chernobyl disaster in 1986, the collaboration with physicians transformed into studies on EPR dosimetry. Although this collaboration did not lead to any joint publications, the method for diagnosing malignant neoplasms of lungs using spin labels and benchtop EPR spectrometer RM-10-40-RD was developed in the N.N. Alexandrov National Cancer Centre of Belarus [35]. It also stimulated to some extent the further development of EPR dosimetry in Belarus [36]. Motoji Ikeya, a Japanese physicist known for his contribution to EPR dating [37] and dosimetry [38,39], visited BSU Industrial Laboratory (mainly transformed to ADANI company by then) in Minsk in the early 1990s to review new EPR equipment and met with one of the authors of this article (V.N. Linev).

6. Conclusions

The diversity of applications resulted in wide range of specialized models of EPR spectrometers. At least nine models were developed and serially produced by the mid-1980s including the RM-6, RM-20P, REM-10, REM-12MFS, REM-10-30, REM-10-31, REM-10-32, REM-10-33, and REM-10-35 models (Figure 6). Three of them, RM-6, REM-10 and REM-10-31, were used in research in geology for paleodosimetry and the detection of deep-lying ore bodies. Their use substantially changed the design of exploration expeditions, allowing completion of the exploration within one season. RM-20P served for technological control of ore enrichment in non-ferrous metallurgy (instead of more time-consuming chemical methods used before). An REM-10-33 EPR spectrometer was developed and passed clinical testing for the early diagnosis of oncological diseases using spin labels. Serial mass production of an REM-10-32 universal EPR spectrometer was launched under the trade name AE-4700 in Lviv (industrial group "Micropribor"). In the late 1980s, AE-4700 was the only commercially produced EPR spectrometer in the USSR.

In addition to the above, in the chemical industry, benchtop EPR spectrometers were used to build an automated control system for preparing a polyisoprene polymerization catalyst; in the non-ferrous-metals industry, to develop methods of quantifying the microcontent of 27 elements; in geology, to elaborate methods for evaluating and searching for deposits of non-ferrous and rare metals; in the oil industry, to provide reliable and safe methods for monitoring the movement of formation water in oil fields; in microelectronics, to introduce non-destructive methods for measuring and certifying magnetic thin films; and in agriculture, to control infestation of grain by pests ([29], pp. 2–3) amongst other applications.

In summary, the impact of benchtop EPR spectrometers (developed in BSU, LETI and "Svetlana") on Soviet economics was assessed as already exceeding some 5 million rubles (i.e., some \$7.5 M) per year by the early 1980s ([29], p. 3). Hence, the series of benchtop EPR spectrometers was awarded several prizes, the most major of which was the prize of the Council of Ministers of the USSR in 1985 (all three groups of developers of benchtop EPR spectrometers, from BSU, LETI and "Svetlana", were awarded). That was one of few prestigious awards in the USSR concerning EPR.

However, along with the industrial applications listed above, these devices were no less important for radiospectroscopy in the USSR. The tragedy for Soviet radiospectroscopy was that the serial production of EPR spectrometers in the USSR was stopped from 1984. Given the difficulty of buying scientific equipment abroad, this had a detrimental effect on science in the USSR. Erlen I. Fedin, who headed the Commission on radio spectroscopy of the Soviet Academy of Sciences and dedicated his life to improving Soviet NMR/EPR instrumentation [5], hoped that this last Soviet EPR developmental project could save radiospectroscopy in the USSR from disaster and provide science and industry with much needed EPR equipment ([29], pp. 162–163). Although it did not save the Soviet EPR instrumentation, the new conditions in the post-Soviet space helped to revive it to some extent in due to the work of the "ADANI" company.

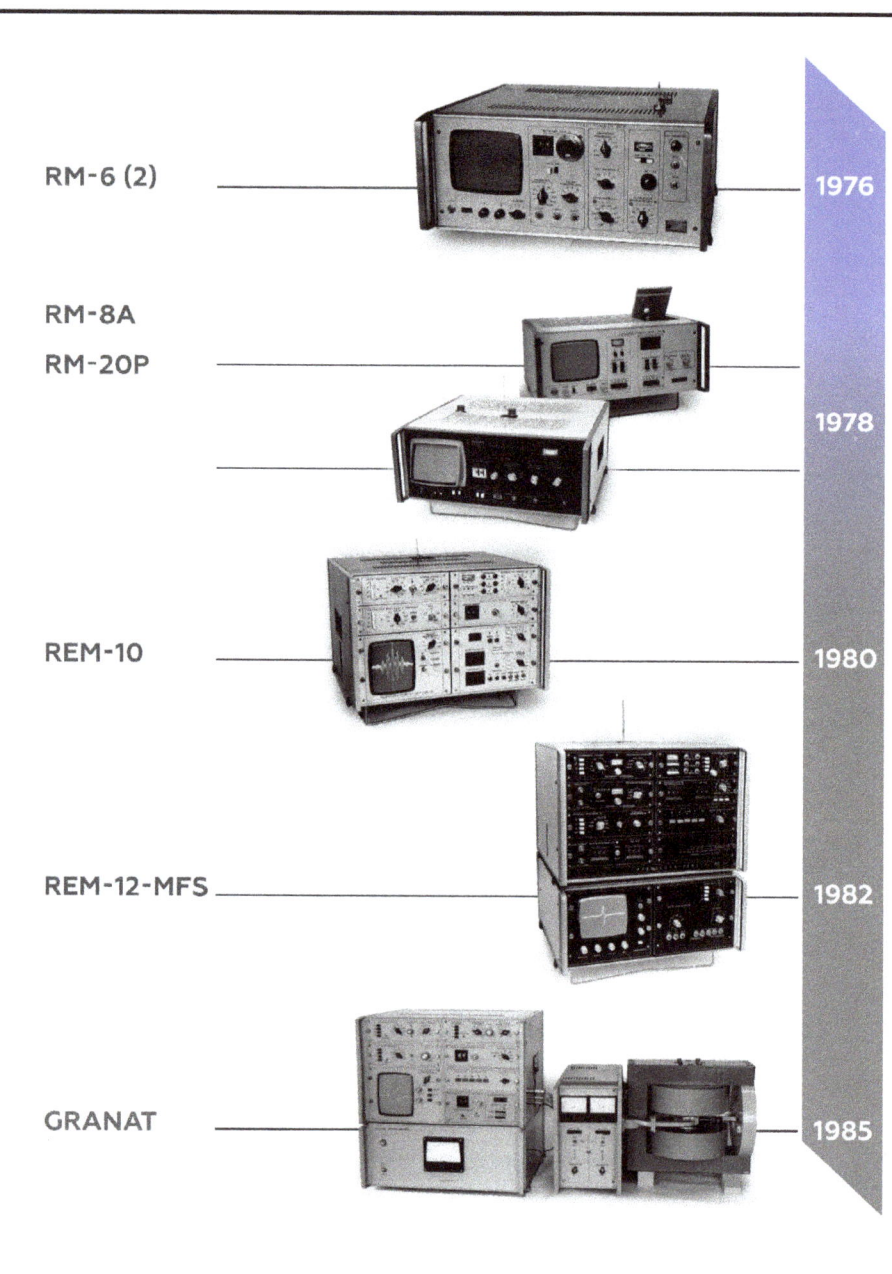

Figure 6. Family of benchtop EPR spectrometers developed and serially produced (in small series) by 1985.

In this paper, we have given a brief history of development of the benchtop EPR spectrometers in the USSR. For the Soviet science, this project was a chance to "revitalize" EPR spectroscopy studies which were being suffocating due to the lack of modern instruments.

However, considering the broader context, this history could cause innovation for the future. In our opinion, benchtop EPR spectrometers production could be a kind of "disruptive innovation" in the sense of Clayton Christensen, i.e., the one that creates a new market albeit initially inferior to the current favorite in terms of operating capabilities [40].

This means we can expect that, in the near future, benchtop EPR spectrometers will conquer the market and, in a sense, will displace the research-grade EPR spectrometers as well as smartphones have displaced the professional photo and video cameras. Already now, there are more than 500 modern benchtop EPR spectrometers operating around the world (Figure 7). New technologies in microelectronics make it possible to find even more compact and low-cost technical solutions for all nodes of the EPR spectrometer than those described above. Although they may not in the immediate future be able to compete with research-grade EPR spectrometers in scientific studies, their cheapness and ease of use can significantly expand the scope of their application in the future, and ongoing innovation will enhance their performance step by step.

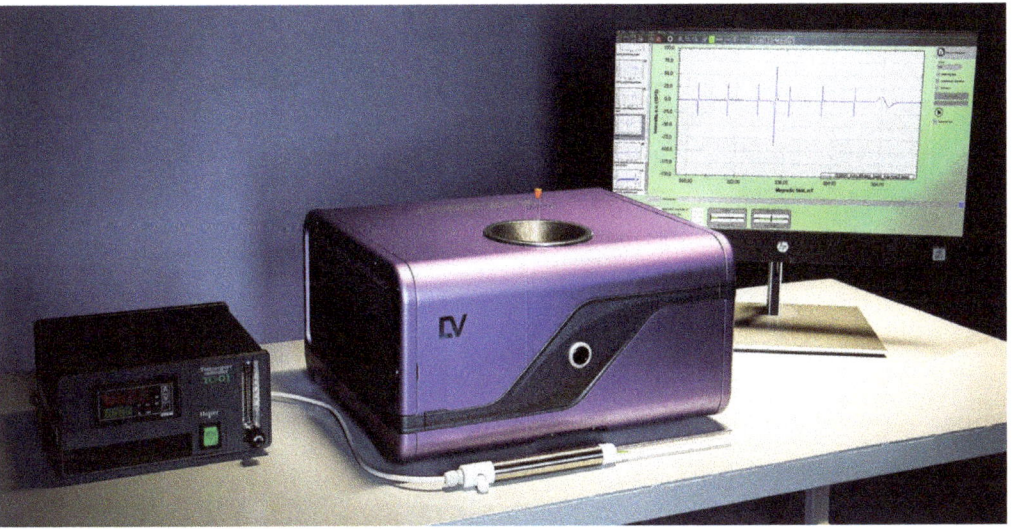

Figure 7. The appearance of a modern benchtop X-band EPR spectrometer (the 2022 model).

Author Contributions: Conceptualization, V.V.P. Writing—original draft preparation, V.V.P. Validation, V.N.L. Writing—review and editing, V.N.L., V.V.P. All authors have read and agreed to the published version of the manuscript.

Funding: This study received no funding.

Institutional Review Board Statement: Not applicable.

Informed Consent Statement: Not applicable.

Data Availability Statement: The data analyzed in this study are available from the authors upon reasonable request.

Conflicts of Interest: V.N. Linev is the LINEV Group UK Founder & CEO.

Sample Availability: Not applicable.

References

1. Hyde, J.S. EPR at Varian: 1954–1974. In *Foundations of Modern EPR*; Eaton, G.R., Eaton, S.R., Salikhov, K.M., Eds.; World Scientific: Singapore; Hackensack, NJ, USA; London, UK; Hong Kong, China, 1998; pp. 695–716. [CrossRef]
2. Salikhov, K.M.; Zavoiskaya, N.E. Zavoisky and the discovery of EPR. *Resonance* **2015**, *20*, 963–968. [CrossRef]
3. Ptushenko, V.V. The unfinished Nobel race of Eugene Zavoisky: To the 75th anniversary of EPR discovery. *Sci. Bull.* **2019**, *64*, 146–148. [CrossRef]
4. Ptushenko, V.V. Chain Initiation. *Her. Russ. Acad. Sci.* **2019**, *89*, 84–90. [CrossRef]
5. Ptushenko, V.V.; Amiton, I.P. To turn the tide in the Soviet scientific instrumentation: In memoriam Erlen I. Fedin (1926–2009). *Struct. Chem.* **2018**, *29*, 1225–1234. [CrossRef]
6. Kessenikh, A.V.; Ptushenko, V.V. *Magnetic Resonance in the Interior of the Century: Biographies and Publications [In Russian: Magnitnyi Rezonans v Interiere veka: Biografii i Publikatsii]*; Fizmatlit: Moscow, Russia, 2019; p. 232. Available online: https://www.rfbr.ru/rffi/ru/books/o_2092943#1 (accessed on 7 September 2022).
7. Ptushenko, V.V.; Zavoiskaya, N.E. EPR in the USSR: The thorny path from birth to biological and chemical applications. *Photosynth. Res.* **2017**, *134*, 133–147. [CrossRef]
8. Shushkevich, S.S. *My Life: The Collapse and Resurrection of the USSR [In Russian: Moia zhizn, Krushenie i Voskreshenie SSSR]*; Rospen: Moscow, Russia, 2012; p. 470.
9. Prokhorov, A.M. Quantum Electronics. *Science* **1965**, *149*, 828–830. [CrossRef]
10. *Personnel File of Alexander A. Manenkov*; Arch. Kazan Phisico-Technical Institute: Kazan, Russia, 1953; F.4, Op.5, d.99.
11. Smolyansky, V.S. *Catalog of Developed Devices [In Russian: Katalog Razrabotannykh Priborov]*; Pavlenko, V.A., Ed.; Nauka: Leningrad, Russia, 1977; p. 159.
12. History of the Foundation of IGEM RAS [In Russian: Istoriia Osnovaniia IGEM RAN]. Electronic Document at the Official Website of IGEM. Available online: http://www.igem.ru/igem_history/history.html (accessed on 16 January 2022).
13. In the Service of Russian Geology (VIMS, 1904–2018) [In Russian: Na Sluzhbe Otechestvennoi Geologii (VIMS, 1904–2018)]. Electronic Document at the Official Website of VIMS. Available online: https://vims-geo.ru/documents/33/VIMS.pdf (accessed on 7 September 2022).
14. Marfunin, A.S.; Bershov, L.V. Paramagnetic centers in feldspars and their possible crystallochemical and petrological significance. *Dokl. Akad. Nauk SSSR* **1970**, *193*, 412–414.
15. Moiseev, B.M.; Rakov, L.T. Formation of Al-centers and E-centers in quartz under action of natural irradiation. *Dokl. Akad. Nauk SSSR* **1975**, *223*, 1215–1217.
16. Kislyakov, Y.M.; Moiseev, B.M.; Rakov, L.T.; Kulagin, E.G. The distribution of E-centres concentration in the minerals of the wall-rocks of uranium deposit [In Russian: Raspredelenie kontsentratsii E-tsentrov v mineralakh rudovmeshchaiushchikh porod uranovogo mestorozhdeniia]. *Geol. Ore Depos. [Russ. Geol. Rudn. Mestorozhdenij]* **1975**, *17*, 86–92.
17. Brief Historical Background. Department of Nuclear Physics. BSU Website [In Russian: Kratkaia istoricheskaia Spravka. Kafedra Iadernoi Fiziki. Internet-Sait BGU]. Available online: http://www.physics.bsu.by/ru/departments/nuclear-physics/history (accessed on 11 November 2018).
18. Bershov, L.V.; Linev, V.H.; Moiseev, B.M.; Orlenev, P.O.; Fursa, E.Y.; Shushkevich, S.S. A specialized spectrometer EPR for field works [In Russian: Spetsializirovannyi spektrometr EPR dlia raboty v polevykh usloviiakh]. *Bull. Acad. Sci. USSR Geol. Ser. [Russ. Izv. Akad. Nauk SSSR Seriya Geol.]* **1981**, 152–155.
19. Linev, V.N.; Muravsky, V.A.; Slepyan, G.Y.; Fursa, Y.Y.; Shushkevich, S.S. Measuring Rectangular Resonator for Electronic Paramagnetic Resonance [In Russian: Izmeritelnyi Priamougolnyi Rezonator dlia Elektronnogo Paramagnitnogo Rezonansa]. Patent SU 968717 A1, 23 October 1982. Available online: https://elibrary.ru/item.asp?id=40289297 (accessed on 7 September 2022).
20. Linev, V.N.; Figurin, V.A.; Fursa, Y.Y.; Shushkevich, S.S. A Method for Stabilizing Resonance Conditions in an Electron Paramagnetic Resonance Radio Spectrometer [In Russian: Sposob Stabilizatsii Rezonansnykh Uslovii v Radiospektrometre Elektronnogo Paramagnitnogo Rezonansa]. Patent SU 1038850 A1, 30 August 1983. Available online: https://elibrary.ru/item.asp?id=40354800 (accessed on 7 September 2022).
21. Linev, V.N.; Lisovsky, V.V.; Muravsky, V.A.; Fursa, Y.Y.; Shushkevich, S.S. A Method for Controlling a Polarizing Magnetic field in Spectrometric Equipment and a Device for Its Implementation [In Russian: Sposob Upravleniia Poliarizuiushchim Magnitnym Polem v Spektrometricheskoi Apparature i Ustroistvo dlia ego Osushchestvleniia]. Patent SU 1000874 A1, 28 February 1983. Available online: https://elibrary.ru/item.asp?id=40099304 (accessed on 7 September 2022).
22. Figurin, V.A.; Linev, V.N.; Lisovsky, V.V. Spectrometric instruments miniaturization of permanent magnets. In Proceedings of the 1st Soviet-Indian Symposium on Actual Problems of Magnetic Resonance Spectroscopy of Inorganic Materials, Dushanbe, Tajikistan; 1982; p. 192.
23. Linev, V.N.; Lisovsky, V.V. Permanent magnet assembly with variable magnetic field for magnetic resonance spectroscopy. In Proceedings of the Magnetic Resonance and Related Phenomena: XXIV Congress AMPERE, Poznan, Poland, 29 August–3 September 1988; p. A–95.
24. Linev, V.N.; Mochalsky, V.B.; Muravsky, V.A.; Fursa, Y.Y. Registration of EPR spectra with variable magnetic field sweep rate [In Russian: Registratsiia spektrov EPR s peremennoi skorostiu razvertki magnitnogo polia]. *Bull. Belarusian State Univ. Ser. 1* **1982**, 8–12.

25. Linev, V.N.; Marshalko, S.V.; Muravsky, V.A.; Fursa, Y.Y.; Shushkevich, S.S. A Method for Isolating Weak Magnetic Resonance Signals and a Device for Its Implementation [In Russian: Sposob Vydeleniia Slabykh Signalov Magnitnogo Rezonansa i Ustroistvo dlia ego Osushchestvleniia]. Patent SU 1100547 A1, 30 June 1984. Available online: https://elibrary.ru/item.asp?id=40230993 (accessed on 7 September 2022).
26. Linev, V.N.; Muravsky, V.A.; Figurin, V.A.; Fursa, Y.Y. A Method for Recording Electron Paramagnetic Resonance Spectra of Anisotropic Substances [In Russian: Sposob Registratsii Spektrov Elektronnogo Paramagnitnogo Rezonansa Anizotropnykh Veshchestv]. Patent SU 1190245 A1, 7 November 1985. Available online: https://elibrary.ru/item.asp?id=40345516 (accessed on 7 September 2022).
27. Bolotskikh, V.G.; Linev, V.N.; Lisovskij, V.V.; Mileshkevich, V.V.; Sokolovskij, V.V.; Fursa, E.Y. Small-sized EPR spectrometer [In Russian: Malogabaritnyi spektrometr electronnogo paramagnitnogo rezonansa]. *Instrum. Exp. Tech. [Russ. Prib. I Tekhnika Ehksperimenta]* **1983**, 244.
28. Kulikovskikh, B.E.; Linev, V.N.; Fursa, Y.Y.; Shushkevich, S.S.; Yanovsky, V.P. Small-Sized Electron Paramagnetic Resonance Spectrometer [In Russian: Malogabaritnyi Spektrometr Elektronnogo Paramagnitnogo Rezonansa]. Patent SU 855460 A1, 15 August 1981. Available online: https://elibrary.ru/item.asp?id=40228003 (accessed on 7 September 2022).
29. RGAE Materials on the works submitted and considered for the award (awarded). 1985. No. 30 instr./mining/chem. Creation of small-sized magnetic resonance equipment and radio spectroscopic methods. *Russ. State Arch. Econ. (RGAE)* **1985**. F. 9480, Op.13, d.2059.
30. Galimov, E.M. Vernadsky Institute of Geochemistry and Analytical Chemistry (GEOKhI): Scientific results in 2011–2015. *Geochem. Int.* **2016**, *54*, 1096–1135. [CrossRef]
31. Zolotov, Y.A.; Petrukhin, O.M.; Nagy, V.Y.; Volodarskii, L.B. Stable free-radical complexing reagents in applications of electron spin resonance to the determination of metals. *Anal. Chim. Acta* **1980**, *115*, 1–23. [CrossRef]
32. Zotov, N.I.; Kozhukhar, A.Y.; Linev, V.N.; Netsvetov, V.I.; Nikolaev, E.I.; Fursa, E.Y. Relation between temperature and mobility of domain boundaries in epitaxially grown ferrite-garnet films [In Russian: Temperaturnaia zavisimost podvizhnosti domennykh granits v epitaksialnykh ferrit-granatovykh plenkakh]. *Phys. Solid State [Russ. Fiz. Tverd. Tela]* **1981**, *23*, 287–289.
33. Kozhukhar, A.Y.; Linev, V.N.; Fursa, Y.Y.; Shagaev, V.V. Method of Research and Non-Destructive Testing of Magnetic Films [In Russian: Sposob Issledovaniia i Nerazrushaiushchego Kontrolia Magnitnykh plenok]. Patent SU 1065750 A1, 7 January 1984. Available online: https://www.elibrary.ru/item.asp?id=40122102 (accessed on 7 September 2022).
34. Revokatov, O.P.; Gangardt, M.G.; Murashko, V.V.; Zhuravlev, A.K. NMR relaxation in human blood serum a cancer diagnostic method. *Rus. J. Biofizika* **1982**, *27*, 336–338.
35. Mashevsky, A.A.; Prokhorova, V.I.; Tsyrus, T.P.; Putyrsky, L.A. Method of Diagnosis of Malignant Neoplasms of the Lungs [In Russian: Sposob Diagnostiki Zlokachestvennykh Novoobrazovanii Legkikh]. Patent RU 2056052 C1, 10 March 1996. Available online: https://elibrary.ru/item.asp?id=38032685 (accessed on 7 September 2022).
36. Ugolev, I.I.; Linev, V.N.; Finin, V.S.; Bogushevich, S.E. Modern condition of the spin-resonance dosimetry. In Proceedings of the Metrology-94: Scientific and Practical Conference, Minsk, Belarus, 1–2 March 1994; pp. 57–58.
37. Ikeya, M. Dating a stalactite by electron paramagnetic resonance. *Nature* **1975**, *255*, 48–50. [CrossRef]
38. Ikeya, M. Paramagnetic alanine molecular radicals in fossil shells and bones. *Naturwissenschaften* **1981**, *68*, 474–475. [CrossRef]
39. Ikeya, M.; Miyajima, J.; Okajima, S. ESR dosimetry for atomic bomb survivors using shell buttons and tooth enamel. *Jpn. J. Appl. Phys.* **1984**, *23*, L697. [CrossRef]
40. Wu, L.; Wang, D.; Evans, J.A. Large teams develop and small teams disrupt science and technology. *Nature* **2019**, *566*, 378–382. [CrossRef] [PubMed]

MDPI
St. Alban-Anlage 66
4052 Basel
Switzerland
www.mdpi.com

Molecules Editorial Office
E-mail: molecules@mdpi.com
www.mdpi.com/journal/molecules

Disclaimer/Publisher's Note: The statements, opinions and data contained in all publications are solely those of the individual author(s) and contributor(s) and not of MDPI and/or the editor(s). MDPI and/or the editor(s) disclaim responsibility for any injury to people or property resulting from any ideas, methods, instructions or products referred to in the content.

www.ingramcontent.com/pod-product-compliance
Lightning Source LLC
LaVergne TN
LVHW070636100526
838202LV00012B/823